A Guide to Agribusiness in Zambia

A Guide to Agribusiness in Zambia

Untapped Opportunities

Felix Tembo

Copyright © 2019 by Felix Tembo.

ISBN: Softcover   978-1-7960-1912-4
            eBook   978-1-7960-1902-5

All rights reserved. No part of this book may be reproduced or transmitted in any form or by any means, electronic or mechanical, including photocopying, recording, or by any information storage and retrieval system, without permission in writing from the copyright owner.

Any people depicted in stock imagery provided by Getty Images are models, and such images are being used for illustrative purposes only.
Certain stock imagery © Getty Images.

Print information available on the last page.

Rev. date: 07/04/2019

To order additional copies of this book, contact:
Xlibris
1-888-795-4274
www.Xlibris.com
Orders@Xlibris.com
788972

CONTENTS

Acknowledgement ix
Foreword xi

Chapter 1 1
   1.1   The climate of Zambia 1
Chapter 2 7
   2.1   Pre-independence agriculture 7
Chapter 3 14
   3.1   Modern agriculture in Zambia 14
   3.2   The agribusiness environment 18
   3.3   Agriculture in the first republic 50
   3.4   Agro market development 53
   3.5   Infrastructure development and agribusiness 59
   3.6   The economy and agriculture 63
Chapter 4 67
   4.1   Politics and agriculture 67
   4.2   Political stability 71
   4.3   Government policies 73
Chapter 5 78
   5.1   Agribusiness and commerce 78
Chapter 6 89
   6.1   Agribusiness in practice 89
   6.1.1 Political will and value addition 90
   6.1.2 The importance of value addition 95
   6.1.3 Agro- enterprises and competitiveness 100
   6.1.4 The panacea for real diversification 104
   6.1.5 Real job creation in agriculture 107
   6.1.6 Sustaining jobs 113

6.1.7	Unlocking the agribusiness potential	117
6.1.8	The ripe agribusiness environment	120
6.1.9	Irrigation development in Zambia	123
6.1.10	Planting a 'bumper crop'	125
6.1.11	Growing tomato is profitable	133
6.1.12	Are tomatoes difficult to grow?	138
6.1.13	Food security and what to grow	141
6.1.14	Diversifying for food security	147
6.1.15	The changing production dynamics	150
6.1.16	Importance of land in agriculture	152
6.1.17	Soybean is 'a money spinner'	157
6.1.18	Onion, the 'Sulphur' mine	169
6.1.19	Impact of rice branding on job creation	171
6.1.20	Investing in seed purification	177
6.1.21	A dead yet viable subsector	180
6.1.22	Aflatoxins in groundnuts	183
6.1.23	Losing money due to 'pops'	186
6.1.24	The foundation for effective crop production	188
6.1.25	Soil organic matter	191
6.1.26	Soil acidity and liming	193
6.1.27	The right time to weed	196
6.1.28	Management of the weed called Striga	199
6.1.29	Do fertilizers destroy the soil?	200
6.1.30	Multi agro-economic facility zones	203
6.1.31	Subsidizing crop marketing	206
6.1.32	Agriculture without subsidies, is it viable?	209
6.1.33	The cotton industry in Zambia	216
6.1.34	Agricultural markets in Zambia	219
6.1.35	Smallholder farmers and crop marketing	222
6.1.36	Smallholder maize pricing nightmare	225
6.1.37	The 'power' of information	227
6.1.38	Markets must guide farmers	230
6.1.39	The well organized dairy sector	232
6.1.40	Input subsidies, have they delivered?	235
6.1.41	Creating agribusiness giants in the sector	239
6.1.42	Market opportunities for cassava	241

6.1.43 Cooperatives can be vehicles for economic development..245
6.1.44 'Zoning' the country 247
6.1.45 Corruption, a thorn in the agriculture sector........249
6.1.46 Follow ethics in crop production 252
6.1.47 The need for competition......................................255
6.1.48 Zambia as the regional hub for agro-investments..257
6.1.49 The effective management of stress in crop production 260
6.1.50 Impact of technology adoption on productivity....262
6.1.51 Financing agriculture for productivity 265
6.1.52 Revamping the agribusiness sector........................270
6.1.53 The private sector as the driver............................273
6.1.54 Covering up for the lost opportunity....................275
6.1.55 Cultivating for 'resilience'.....................................278
6.1.56 Is agribusiness on the right track? 280
6.1.57 Is agriculture profitable in Zambia? 283
6.1.58 The ban on cooking oil importation....................285
6.1.59 Nematodes; a menace to crops.............................287
6.1.60 Preserving 'money' in the harvest.........................289
6.1.61 Fall army worms in Zambia..................................292
6.1.62 Should we invest in marketing or productivity improvement?..295
6.1.63 The high prices of mealie-meal............................298
6.1.64 With five acres, what can one grow? 301
6.1.65 What a huge potential for the livestock sector! 303
6.1.66 Potato, an alternative crop to tomato...................308
6.1.67 The sustainable management of natural resources for tourism 311

Acronyms...317
About The Author..321
About The Book 323

ACKNOWLEDGEMENT

In this book, I want to share with my fellow countrymen and women how I have viewed agricultural development in Zambia from the time we got independence to date. I want to identify some enablers that have contributed to its development and some of the constraints that I perceive to have hindered its further development. I have endeavored to highlight some of the available opportunities that one can find in the country as well as identify strides made to make markets function. I should state that the views expressed in this book are from a personal perspective except in instances where I have quoted some scholars. In addition, I have decided to publish some of the articles I contributed to one of the daily tabloid newspapers of Zambia in a column I was writing from 2012 to early 2017 in a feature column called *Agribusiness Focus*. This is because of the many letters I received from my readers of the column asking me to document the articles in form of a book. I am also indebted to my colleagues who encouraged me to put my ideas to paper so that many people can benefit from. It is not possible to mention everyone but suffice to say I received so much encouraging messages from the readers, notable amongst many being Daniel, Alfred, Sharma, Mulenga, Aggie including Patel from Pakistan who constantly emailed me to find out whether the book was ready.

    I am highly indebted to my family who understood and where patient in allowing me a lot of time and space to be alone in silence despite many demanding challenges at home. They were so supportive especially my wife and the children. I would like to give special thanks to Richmond who did the proof reading and Dr Eike Hupe for agreeing to write the foreword to this book. The book indeed is dedicated to the smallholder farmers of Southern Africa and Zambia in particular who have managed to feed this country for a very long time under very difficult operating business environment.

FOREWORD

The first time I came across some of the original content of this book was in 2014. I was about to drive from Lusaka to Chipata in Zambia, a distance of six hundred kilometers and stopped to buy a local newspaper. At the time, I had just come to Zambia to take up my new role and I was trying to understand the context of Zambia's agriculture. As I perused through the newspaper I saw a highly interesting article written by this author, on agricultural practices in Zambia. Coming from Europe where agriculture is different from this part of the world, it really gave me great insight into Zambia's agriculture and the opportunities found in this country; from then onwards I periodically read these articles.

A few years later, I was fortunate enough to have asked Felix to join my team and he gladly joined us as the regional agronomist, responsible for driving business development in nine Southern African countries. Sometime later I asked Felix how his writing was going on and encouraged him to consolidate his articles and publish them as a book. And after having seen the final result, it was indeed a successful undertaking which I believe will help a lot of farmers in Zambia and those that would want to invest in the agricultural sector in Zambia.

With this book, *"Agribusiness Environment in Zambia; Untapped Opportunities"*, Felix is supporting the development of the agricultural sector in Zambia in particular and Southern African region in general. This is a sector which has shown consistent growth over the years and which will continue to have great potential in the years to come. If we just take a look at maize, the main crop grown in Zambia by the smallholder farmers, we see that today the average maize yield for smallholders in Zambia is a little above two tons per hectare as compared to a couple of years back when it was at one and half tons per hectare and below. With the right trainings through extension, adoption of technologies and better farming tools, I am convinced that smallholder farming will continue to thrive and will achieve improved yields of up to five tons per hectare and more, thus more than doubling the contribution to the country's gross domestic product. The potential for growth is certainly significant with the over than a million hectares that are cultivated by the smallholder farmers. Additionally, Zambia's commercial farming sector has great growth potential in the region. The quality and availability of land that can be cultivated is certainly outstanding and of its own on the continent. Let's take sugarcane as an example. On average a field in Zambia you can easily achieve more than hundred tons per hectare – this compares favorably with Brazil where yields are averaging eighty tons per hectare.

I have no doubts that with the adoption of various new technologies - as highlighted in this book - the competitiveness of agricultural products from Zambia and the region will further increase. This indeed will be a springbok for Zambia's agricultural development which ultimately will lead to economic development of the people of Zambia and the country as a whole.

Therefore, when Felix asked me to write the foreword to his book, I felt honored – having had the chance to get to know him as a highly gifted agronomist, a hard-working individual, and an excellent team player. I have no doubt that his book will help to further develop the agriculture sector in Zambia, and thus be an important contribution to the wellbeing of farmers in Zambian in particular and the sub region. I therefore encourage those that want to better their agriculture and those with intentions of investing in Zambia's agricultural sector to read this book.

Eike Hupe (Dr).

Managing Director
BASF Zambia Limited 2017

CHAPTER 1

The climate of Zambia

Zambia is a land-locked country found in South Central Africa. It shares borders with eight countries, namely: Angola to the west, the Democratic Republic of Congo (DRC) to the north, Tanzania to the northeast, Malawi to the east, Mozambique to the southeast, Zimbabwe and Botswana to the south and Namibia to the southwest.

Zambia's potential arable land covers forty-seven percent of the country's total landmass but only fifteen percent of this is under cultivation. Cropped land is estimated at 7.08 percent, of which permanent crops occupy less than one percent (0.03 percent) and others 92.9 percent as of the 2001 and 2005 estimates. In this book, we shall not dwell so much into the history of how agriculture started in this part of Africa because that is beyond the scope. The focus will be to highlight the agricultural development from the time just before independence to date in a summary, and elaborate the potential that the country has. This may allow us to understand why the country is where it is today. By so doing, it would be easy to know how we need to forge ahead with the view to improving the agribusiness industries and the sector as a whole. Zambia has a favourable climate and it is divided into three (3) main agro-ecological zones called regions; region I, region II which has two sub regions II*a* and II*b* and region III. These are classified according to the amount of rainfall received and the type of vegetation predominantly found in those areas.

The country has one main rainy season which runs from November to April of every year. However, with the advent of climate change, the onset of the rainy season has shifted and it has been delaying. The rainy season is now starting between late November and mid-December and ending in early March. This has been attributed to the effects of global warming which has seen the global temperatures rising. Ultimately, this has not only led to changes in the rainfall patterns but the intensities as well. The agro-ecological classification is based on the amount of rainfall received which has a bearing on the predominant vegetation found in each agroecological zone. Region one is the area that receives rainfall between 400mm to 800mm per year. These are areas that are found mainly in the valleys; predominantly in many parts of southern Zambia such as the Zambezi valley, part of the extension of the Kalahari sands that stretches into the southern parts of Western province. This region also includes the Luangwa valley and some southern parts of Eastern province. It is predominantly dry with sturdy vegetation and is sometimes called the Luangwa – Zambezi Basin.

The second agro-ecological zone, called region II is the area running across the middle parts of the country and lying north of region I and south of region III. This region covers the northern districts of southern province, all of central province, Eastern and the southern parts of Muchinga provinces. The region receives rainfall ranging from 800mm to 1000mm, and this region extends into Western province and the southern parts of North Western province where it is classified as region IIb. The remaining areas which are mostly in the northern parts of the country covering the northern districts of Central province, Copperbelt, Luapula, North western and Northern provinces are classified as Region III. This region predominantly receives rainfall above 1000mm annually.

The classification of the county according to the rainfall patterns is important in that it dictates what crops can be grown in each of the regions. This is also true for the soil types found in each region. The amount of rains received also determines the predominant type of soils found in those regions. For instance, region III is characterized by acidic soils (low in pH) and highly weathered such as the oxisols. This is even true for the type of vegetation although the largest extent of Zambia is predominantly found with savannah type of vegetation. It would be important to state that the country is endowed with abundance of water from many rivers, lakes and streams that are found in all parts of the country. The major lakes are Bangweulu and Mweru in Luapula province; Lake Tanganyika and Mweru Wantipa in Northern Province; Lake Kariba, which is the second largest man-made lake in the world found in Southern province. We also have so many dam like lakes such as the Itezhi Tezhi in Namwala, the Kafue lagoon, lake Chila in Northern province and so many other water bodies such as Nampamba in Ndola rural from which Mpongwe Estates draws its water for irrigation. Of the major rivers, we have the famous Zambezi River which runs through five provinces and borders with five countries namely; Northwestern, Western, Southern, Lusaka, Eastern provinces and the countries are Zambia, Angola, Namibia, Botswana, Zimbabwe and Mozambique. The source of this great river is found in Zambia at a place called Kalene hills in North-western province near the border with D.R. Congo. The source of Zambezi river is a very big tourist attraction which is visited by thousands of tourists in a year. At some point, this river runs parallel with Kafue river whose source is on the Copperbelt province. It also passes through Central, Southern and Lusaka provinces before emptying its waters into the mighty Zambezi river. The other great river is the Luangwa River which has its source in the Northern Province and runs through the valley bordering the Muchinga escarpment through Eastern province before offloading its water into Zambezi River at the border with Mozambique, Zambia and Zimbabwe. Chambeshi River is another one which starts in the Northern Province and runs into Lake Bangweulu on the northern side and emerges from the southern end of Lake Bangweulu as Luapula River and runs up north forming a natural boundary with the D.R. Congo before finally emptying its water into Lake Mweru. Besides these great rivers, we have other important streams and rivers like the Mulungushi, Luongo, Lunga, Lungwebungu, Lukusashi and many others who empties their waters in the big rivers. Most of these rivers and streams sources are the aquifer in the Congo Basin. It is gratifying to note that forty percent of the water found in the SADC region (Zambia, Zimbabwe, Angola, Malawi, Mozambique, South Africa, Namibia, Botswana, Swaziland, Lesotho) is in Zambia. This is a great opportunity and a comparative advantage that the country has over other countries in the region.

Zambia's climate is one that is favourable for agricultural production especially for field crops, vegetables as well as citrus fruits and rearing of livestock. The country has three distinct seasons. The rainy season, which is the period from December to April, is characterized by high temperatures of over 25oC with high humidity and cloud cover (rainy clouds). It is normally wet and can get very warm up to as high as 40^0C and humid at times. This is the main production season especially for crops such as maize, sorghum, millet, cassava, sweet potatoes, soybeans and many other crops. This is followed by a cold season (winter) which runs from May to August. The temperatures are generally low, and sometimes getting very cold. The temperatures drop to as low as 10^0C while the nights are very cold and in some places like Mpika, Katete, Kalomo, and Sesheke temperatures can drop to as low as 4^0C and sometimes below zero. Towards the end of the cold season, the days are windy and dusty. However, the average day temperatures fall within the range of 15^0C to 22^0C. This is the main season for wheat production. The hot season is the period from September to November. This is generally dry and hot. Temperatures range between 30^0C to 38^0C, with valley temperatures soaring as high as 42^0C. In the valleys and some parts of the country like Western and Southern provinces especially Livingstone, temperatures can rise as high as 45^0C. The sky is clear, dry while the nights are humid and hot. Most of the vegetables are produced during the hot season using various irrigation systems.

Zambia was a British protectorate and it got its independence in 1964. It is one of the few countries in Southern Africa that has experienced peace since its independence. The country is rich with many cultural groupings and has over seventy-three ethnic groupings. In the Northern province, there are found the Bemba, Bisa, Lungu, Mambwe, Namwanga; Central province has the Lenje, Swaka, Lala, Sala, Kaonde; Coperbelt has the Lamba, Lima; Northwestern has the Luvale, Lunda, Kaonde, Mbunda, Kachokwe, Luchazi; Western province has the Lozi, Makoma, Kwangwa, Nkoya; Lusaka has the Goba, Lenje, Soli, Sala; Eastern province has the Ngoni, Chewa, Senga, Nsenga, Kunda, Tumbuka and Southern province has the Tonga, Toka-Leya, Ila while Luapula province has the Ushi, Shila, Lunda, Chishinga. However, these are not the only tribes found in Zambia like earlier mentioned that there are seventy-three (73) tribes or dialects in total.

Each tribe in the country plays a very important part in the history of the country as well as agriculture development is concerned. For instance, the Tonga's and the Lozi's are good herdsmen and they rear about sixty percent of the cattle population found in the country, the Bemba's are good at fishing and cultivation of crops like cassava and millet while the Lamba's are generally good hunters, and so on. These tribes have developed traditional cousinship out of activities that were done earlier in their settlement dates; for instance, the Bemba's fought the Ngoni for land in the 1630s, the Lozi's and Tonga's used to raid each other for cattle and women, it is claimed that the Lamba's showed the Lozi's how to operate the muzzle loader which was the first types of guns used then for both hunting and defending the chiefdoms. It should however, be mentioned that the Tonga's are amongst the first tribes to have adopted commercial farming in Zambia. They are the first tribe to have first embraced agriculture as a business after learning it fom the white settlers. However, this is not the case currently because other tribes have now realized the importance of agriculture as a livelihood and its contribution to job creations, income generation as well as the country's economic development. Agriculture has evolved through different stages from agriculture in pre independence to what it is today. This sector is further destined for growth as the farmers adopts new farming techniques and technologies. There are also some experienced farmers that have come to settle in the country from other neighbouring countries like Zimbabwe and South Africa, as well as from Europe where they can't find new land to expand their production. The country has also experienced the Indian community getting out of their traditional business of trading into agriculture as well as the Chinese because of the favourite climate, policies as well as the ease of doing business in the country, and the accommodative settlement policy of the government as well as the welcoming culture of the people. This could be the reason that the country has been a harven of refugees from Congo D.R., Burundi, Rwanda, Angola, Mozambique, Somali and Sudan for a very long time. Some have even acquired Zambian citizenship.

CHAPTER 2

Pre-independence agriculture

Agriculture in Zambia dates as far back as 1600s when the first settlers came into this part of Africa. However, not more than four centuries ago, our fore fathers depended on hunting and collection of wild fruits for their survival. One would argue that probably it was due to the abundance of natural resources and foods (wild animals, tubers and fruits) that led to this nomadic life, but civilization also plays a key role. Literally every part of Zambia had animals and one did not need to cultivate for one to get food. At some point, there should have been more wild animals than people in this country. This was coupled with a lot of fruits, roots and other foodstuffs from the wild that they easily collected for food. As the population increased and people became more civilised, food became scarce. The coming of the white missionaries made agriculture or farming inevitable. It was slowly being introduced as the people's livelihoods as well as for economic benefits.

Recorded civilisation in this part of the world started between the 16th and 19th centuries as that is when most books of history record organised use of tools made out of iron, and the period is called the Iron Age. Most of the settlements then were according to traditional rulers called chiefs, hence, their communities were called or arranged according to the chiefdoms. These have continued to date and we have notable chiefdoms like the paramount Mwata Kazembe Chiefdom in the north with its headquarters at a place called Mwansabombwe in

Luapula. According to tales told by our senior citizens, this chiefdom was a break away from the Mwatayamvo chiefdom in Congo D.R. called Zaire then. The other chiefdom is that found at the current town called Mungwi with the paramaount chief Chitimukulu as the leader. These are the Bemba speaking people who again are believed to have come from Congo D.R. These might have come slightly earlier than Lunda people with paramount chief Mwatakazembe. Down south east are two paramount chiefs; chief of the Chewa speaking people that settled in the Eastern province with its headquarters at Katete and their chief is called paramount chief Gawa Undi and another one of the Ngoni people that settled in Chipata called paramount chief Mpezeni of the Ngoni tribe. These two paramount chiefs are said to have come from South Africa and are believed to have been running away from the great king Shaka Zulu. In the western part of the country is found King Lewanika, the king of the Lozi people. He is the only traditional leader in Zambia that carries the title of king. They are also believed to have come from the south as well through Namibia and settled along the plains of the great Zambezi river. Its ruler is called the King Litunga, and this kingdom has two capitals: one at Lealui where he stays during the dry season when the waters of the plains have receeded, and the upland at Limulunga where he stays during the rainy season when the plains are flooded. This is near the modern town called Mongu. The movement from Lealui to Limulunga in the wet season takes place annually and the ceremony is called the Kuomboka ceremony. It is an annual festivity that is still celebrated todate around March/April of every year. It is a very rich ceremony which attracts people from all over the world. It should also be mentioned that each of the other four chiefs have annual ceremonies which are equally important and recognized globally. For instance, the two ceremonies that take place in Chipata (February) and Katete (August) attracts chiefs who are subjects of the two paramount chiefs from Malawi and Mozambique. The Lamba's and other tribes that settled in the central parts of Zambia and Copper belt province have a very blurred origin. Some people claim that the Lamba's together with the Bantu Botatwe (Tonga, Lenje and Ila) could have been the first ones to have come from the Lunda-Luba kingdom.

Normally, these treks were as a result tribal wars that ensured during that time. However, I am not an historian but my go at history was to try and show how and where the settlements took place. It is important to understand how the tribes were distributed in the modern Zambia for us to appreciate well the agricultural development that followed thereafter. For instance, you will appreciate that the Lozi's and the Tonga's are tribal cousins because these people were involved in fierce tribal raids or fighting for grazing land and one of these two tribes would steal the other's animals and take away the women as their wives. It is the reason that the Tonga's and the Lozi's hold over sixty percent of Zambia's cattle population. Their keeping of cattle hails back to pre-colonial times and it is 'embeded' in their DNA. The determination of how rich one is by the number of cattle one has. Certain individuals especially with the Ilas can have as many as ten thousand cattle.

Books of history indicate that the first recorded visits by Europeans to Zambia were the Portuguese around 1700 and they came from Mozambique to Feira, morden day called Luangwa. Their mission is claimed to be that they were looking for a route to the east coast to trade with the chiefdoms located inland, mainly in ivory and slaves. As earlier stated, Zambia was colonized by the British, and the first known and recorded Briton to set foot on the Zambian soil was David Livingstone. He is believed to have started his famous journey to explore the upper Zambezi river. In the process he became the first European to see Mosi-oa-Tunya falls around 1850; the waterfalls on the Zambezi river which he named after Queen Victoria of England. He undertook two more exploration journeys. This missionary, David Livingstone later died in Zambia at Chitambo in Serenje on his third journey. The falls are a famous tourism site and they are near a Zambian town named after him called Livingstone town. The Victoria Falls is amongst the seven wonders of the world; it is found in Zambia and part of it is shared with Zimbabwe. The best view of it is on the Zambian side.

In pre-independence times, there is less to be discussed about agriculture in general. Agriculture was mainly done as a way of life. As a young boy while sitting around a fire in the evenings, my grandmother would tell us stories of how good a hunter her husband was. He had married four wives; he had a wife in all the chiefdoms he visited on his hunting escapades. My grandmother was the second wife and she had four children with him; one son and three daughters. He was a good elephant hunter who rarely went to the field to cultivate. The cultivation was left to women who generally cultivated around anthills. When I tried to inquire as to why they cultivated mostly on anthills, my great parents could not give me a clear scientific answer apart from stating that it was easy to cultivate on anthills than flat land because of less weed infestation and the fertility around the anthills. I know there was more to it than meet the eye; there are more scientific explanations to this than the reasons she gave. It wasn't any type of anthill that they would cultivate on but they would meticulously choose the best clayey antihills to cultivate on. If we were to get the details of how the anthills they cultivated on were formed, one would understand why. Anthills are formed by ants that bring to the surface soils that they make from the debris of plant material that they chew in the process of extracting their food. This mechanism tends to bring to the surface the nutrients that could have been deposited deep down the sub-surface, deeper for the plants or crops to access under normal circumstance. A few years back, I learnt of some farmers especially in Southern province that were digging out soils from anthills and spreading it onto their fields. I have also seen this activity being practiced in Zimbabwe and Malawi, especially now that the fertilizer prices have soared beyond reach for many resource poor farmers. This trend of cultivating on anthills has continued to this date especially amongst the Lamba's and many others that may not afford to buy fertilizers. Mostly the crops grown on anthills were sorghum, pumpkins, maize and to some extent pearl and finger millet.

The other type of cultivation which was popular is the one practiced by some people of Northern and Muchinga Provinces called the *Chitemene system*. This is a cultivation system in which small branches of a tree are chopped off and heaped together before being set on fire after they have dried up. Analysing the science behind this practice, it was mostly done in region three which has soils that are predominantly acidic because of high rainfall. By burning the branches, this tended to neutralize the soil acidity from the potash that resulted from the burning. This was generally done on small pieces of land of about a Lima (quarter hectare) to two Lima (0.5ha). One person could cultivate several of the *chitemene* plurally known in the local language as *ifiteme*. After a season or two of cultivating on it, the piece of land is left to fallow for several years and this allows

Picture showing a newly prepared field through Chitemene system

the branches on the remaining tree stumps to regrow or rejuvenate. In short, this is a shifting cultivation type of agriculture. This type of cultivation is now discouraged, as it is believed to be destructive and has negative impacts on the environment as they promote deforestation when not well managed. Notwithstanding the fact that a few people still practice this type of cultivation. Normally, millet and sorghum were the main crops grown on the *chitemene* especially in the first season and they could plant them with cassava in the second or subsequent seasons. There are several other farming systems that are practiced in different parts of the country such as the *fundikila* and so on. The choice of the type of farming system depends on the tribe and type of vegetation as well as crop to be grown.

On the other side of the country, down south and west, mostly they were and are still herders of cattle as earlier stated and grew sorghum and maize with some adopting tobacco and cotton as commercial crops. It is believed that the Tonga's could have been the first indigenous Zambians to practice commercial agriculture. This is attributed to the fact that the first white settlers from Europe through South Africa and Zimbabwe settled along the line of rail from Livingstone, Mazabuka up to Kabwe in central province. The Tonga's started adopting the agricultural practices presumably copied from these white settlers. This, with cultivation of tobacco could be the reasons why most parts of Southern province don't have trees to date. Bigger chunks of land were opened up for cultivation of tobacco as a cash crop and in the process trees were cut down to be used for curing of the crop. This led to deforestation of huge lands. Vegetables and other crops were also grown mostly for sale on the Copperbelt where mines had been opened, which created a big market opportunity for the people that had flocked to seek employment.

Agriculture is a very important sector in Zambia and it contributes above twenty percent to the country's gross domestic product (GDP). It is the sector that employs much of the population of the country; both the old and the young, indirectly or directly especially in rural areas. It employs an estimated forty-eight percent (48%) of the country's workforce. Zambia has a landmass of about seventy-five million hectares (75mio ha) of which about fifty-eight percent (58%) or forty-two (42) million hectares is arable and fourteen percent or 10.5million hectares of arable land is cultivated. Before independence, Zambia's population was just slightly over three million people. However, as the urban areas became more industrialized, most of the people moved from the rural areas to go and work in the towns where mines were being established as well as other supporting industries that needed labour. To some extent, most of the people that remained in the rural areas were either too old or young to work while the energetic and productive youthful rural dwellers trekked to go and seek employment. This had a negative effect on agricultural development in the country as real agriculture was left to foreigners, mostly white farmers and retirees that went to settle back on their traditional lands in their villages. Not too long ago, this had been the trend where the retirees go back and set up farms. However, of late we have seen the emergence of young Zambians taking up agriculture quite seriously as businesses. This has been necessitated by lack of white collar jobs and the ease with which one can go into faming as opposed to establishing other businesses which needs a lot of initial capital injection especially finances.

In a nutshell, pre-independence agriculture was more of primitive in nature than what it has evolved to date. In terms of livestock, most farmers kept small livestock such as chickens, goats and pigs while the Tonga, Ngoni and Lozi mostly were the ones that reared cattle. None of the farmers kept fish because it was a resource which was so abundantly found in natural water bodies. As a source of protein, they mostly depended on hunting of wild animals to supplement the proteins from fish. Therefore, agriculture in pre independence era was not taken so seriously as we have seen it today. Notwithstanding the fact that there is still much to be done to develop agriculture in Zambia and it is one sector which is so promising with abundance in opportunities. There is ernomous potential in this sector and people that are looking for investments should consider agriculture because it is a growing industry.

CHAPTER 3

Modern agriculture in Zambia

Zambia got her independence from the British government in October 1964. At independence, the population of Zambia was just over three million people. The areas which were developed are those along the line of rail stretching from Livingstone in Southern province to Kitwe on the Copperbelt. These had basic infrastructure such as roads, postal services, housing and so on. The major economic activities and foreign exchange earners were mining of copper, cobalt, lead and zinc which was the main industry. This however, has continued to be the major foreign exchange (forex) earner in Zambia to date. Most of the people moved from the rural parts of Zambia to go and work in the mines, mostly in the towns on the Copperbelt such as Luanshya, Ndola, Kitwe, Kabwe, Chililabombwe, Mufulira, Chingola and Maamba. The mines attracted people as far as Malawi, Tanzania and Zimbabwe. The only towns that were mining towns outside the Copperbelt (*ku migodi*) were Kabwe which had the first mine to be opened in Zambia in 1921, where they were mining zinc, copper and lead, and Mamba in the Southern province where they were getting coal and are still mining it which was used as a source of energy in the copper mines. Later, Nampundwe mine was added to the main mining towns.

People moved from all parts of Zambia to go and seek work in the mines. It is the reason that even *Chachacha,* the fight for independence, when it got to that part of the country, it became so fierce because of the cosmopolitan nature of the people found on the mines. We had great names like the late Lawrence Katilungu, the Walamba's of this world and many more who were in the workers union that fought for the workers rights. This proceeded to the living legends like the Kaunda, Wina, late Nkumbula, late Kapwepwe and many more not to forget the famous Mama Julia Chikamoneka, the late Mama Kaunda, Mama Salome Kapwepwe and others. The moving of large population of people into the Copperbelt created opportunities for markets of vegetables and grains such as maize which is the staple food crop in Zambia. Therefore, even commercial agricultural activities were centred on the demand from the mines.

Nonetheless, at present agriculture has spread even into non-mining areas and the biggest driver has been good policies brought about by the government. Dr Kaunda and his comrades, in spite of their humble education had a vision beyond mining. They created several farming blocks away from the mining towns such as the tobacco schemes in Kaoma, Chibwe in Kabwe, Serenje, Chipata and other parts of the country.

The government further created commercial farming blocks in Mkushi which currently is the largest in the country with over sixty commercial farms cultivating over 2,000 hectares under tobacco, 8,000 hectares under maize both commercial and seed maize and 20,000 hectares under wheat and over 30,000 hectares under soybeans. This is the block with the highest concentration of commercial farms in one area. Further, the government started the Mpongwe Development Scheme in Ndola rural which was a wheat scheme that later was run by Commonwealth Development Corporation (CDC) and in collaborations with the Mines. This was from 1982 until early 2000 when it was sold to ZamBeef and Agrivision groups. The estate had about five farms cultivating maize, wheat, soybeans, coffee and cattle ranching farm in Kampemba. This was a good training ground for fresh graduates from agricultural institutions and I happen to have done my industrial attachment at this farm on two occasions. It has produced some of the finest farmers and farm managers this country has ever seen such as some of them found in Mkushi, Chisamba, Southern province and Mpongwe. Mpongwe Development Company (MDC) as it was called then, was a marvel of a farm! The estate followed both a vertical and horizontal integration of its products in that it had a milling company based in Kitwe where mealie-meal and flour products were produced besides other products. The advantage of such schemes is that even small scale farmers surrounding such establishment, learnt good agricultural practices (GAP) and most of them have adopted such practices and have graduated into semi commercial farmers today.

As earlier mentioned, Kaunda's government may have its own shortcomings but generally, it meant well for Zambia and the sub region. His government with the help of other partners went ahead to establish vegetable growing farm for the University of Zambia called York Farms (though the shares have now been sold off, and the university may be a minor shareholder). This is a farm located along Kafue road behind Andrews Motel in Lusaka. This is one of the model farms that practice cultivation on a limited piece of land all year round. They grow export vegetables such as green beans, baby marrow, baby corn, carrots, peas, onions and flowers. You will later read in this book about an article that talks about the good agricultural practices done at this farm. While studying at University of Zambia, I did my second and third year attachments on this great farm. I wondered then why they did not have soil compaction due to the heavy equipment they use for different operations and the fact that heavy machinery and equipment run over the land on a daily basis for the whole year. I conducted a study for my final year thesis on the effectiveness of the land management practices done at the farm in avoiding soil compaction, maintaining soil organic matter and other parameters. The results were amazing.

Generally, farms around Lusaka and Chisamba are the ones involved in production of export vegetables and flowers because of their proximity to international airport at Kenneth Kaunda International Airport (KKIA). There were other enterprises established in other parts of the country, like the banana scheme at Mwense district called Mununshi (though it is no longer operational), the Nakambala sugar estates with its out-grower scheme in Mazabuka. This was the single source for sugar in the country for the industries and for export to EU and East Africa countries such as Burundi and Rwanda through the lake Tanganyika. Of late, there has been other private companies that have set up sugar estates such as the Consolidated Farming in Nampundwe where they built a factory and are growing sugarcane and processing it into sugar. There is also the Kalungwishi in Northern Province, and many other plantations that are in the process of being established in Luapula called Mansa Sugar and Luena as well as one in Western province. Suffice to say that there are still great opportunities for more sugar estates that can be established in the country as the market for sugar is huge in the country and the region. When industries get established, they will be using this raw material in various operations to come up with food stuffs. The bioethanol is another opportunity that the sugar industry can take advantage of and tap into that market.

On the other hand, there are more than 2.2million small scale farmers growing different types of crops in Zambia. These have different land holdings ranging from as small as half (0.5) hectare to about fifty hectares. Most of them are involved in production of maize, cotton, sunflower, sweet potatoes, cassava, groundnuts, and different types of vegetables. Some of the farmers also keep livestock such as cattle, pigs, goats and a few keep sheep. Others keep birds such as chickens for both meat and eggs, ducks, quails, guinea fowls and pigeons especially those near the main markets such as Lusaka and Kitwe. In spite of all this production, the country still imports some poultry products from South Africa. Of late, we have seen some of these smallholder farmers venturing into cash crops such as tobacco, cotton and soybeans because of the opportunities that exists in these subsectors. It is an industry in its infancy when viewed from the the productivity end but with the enormous potential to be the main stay for Zambians and the economy for a long time. This is a sector which needs huge investments in technology, financing and machinery for it to reach the levels of Brazil, Argentina and USA. However, the development of the sector has been dogged with so many problems, which could have led to low productivity and may be it is important that we discuss the agriculture environment at this point.

The agribusiness environment

Zambia is a country endowed with abundance natural resources such as water, land with good fertile soils, relatively a good climate, a young population and the country has had a stable political environment ever since it got its independence. From the time the country got its independence in 1964, it is one of the few countries that have enjoyed remarkable peace in the region. This is a good ingredient for not only agriculture investments but other investments and supporting industries as well. However, it should be noted that peace and political stability of a country is just but some of the factors that can promote investments and development in a country. There are several other factors which must be considered, such as good policies, infrastructure, markets and many others which supports agribusiness development in general.

Economists argue that the market is one of the biggest factors which determine the investments in a sector and country. I tend to agree with them in that though you may have a peaceful environment, you can only develop if you can sell the goods and services that you are producing or offering. The determinant of the market for a country is its population and the level of economic development and integration with other countries. In Zambia, after almost fifty-four years of independence the population is slightly over seventeen (17) million people only. This does not make economic sense to provide a sustainable market for the goods and services that can be churned from the potential that the country has. In economic terms, some scholars suggest that the minimum population that can warrant a market is thirty to sixty million people. I strongly agree with them because if you look at the investments of multinational companies in the region, most of them rush to establish their regional officers either in South Africa or Kenya even though they could be targeting the Zambian raw materials. It is very difficult to justify the reason in the mining companies to go and establish their offices in Durban or Pretoria when their target commodity is copper from Zambia and D.R. Congo. The only sane reasons are that our population is small and the level of supporting infrastructural development such as the roads, communication, rails, electricity, financial institutions and economic intergration are in the infancy. One could argue with the facts that why don't multinational companies then establish their offices in D.R. Congo because that country's population is about five times that of Zambia and it has other precious minerals that South Africa or Kenya don't have. The logic answer is that D.R. Congo has seen a lot of wars since the post Mobutu regime the former leader of that country who later died in exile. Additionally, the infrastructure development of that country is worse off than even the basic ones found in Zambia. Up to date, there are still wars going on in Congo especially in the area around Goma which borders with Rwanda.

There are a number of factors that are important to the development of a country's agriculture sector as well as other sectors too. Some of which are climate of the country, policy and regulatory einvironment, access to sustainable markets for the produce, appropriate technologies, affordable finance, effective extension services and education, sociocultural conditions amongst others. However, Zambia still enjoys a good opportunity to tap into the regional market because it is a member of the regional organisations such as COMESA and SADC. To understand the agribusiness environment in Zambia, it is prudent to discuss some of these major factors that supports the sector individually. Out of the main important factors, let us discuss the following:

Power Generation and Supply – for reasonable economic development to take place in the agribusiness sector such as agro-manufacturing industries, there need to be adequate power generation and supply. Zambia's main electricity source is hydro, and it has the Kariba as the main power source, which was built to supply the newly established mines on the Copperbelt then. The infrastructure was built some time just before independence around 1957. That has been the main source of power that is used in the mines, on farms and industries including households for domestic usage. Agribusiness development needs a lot of power to be used to irrigate fields for instance. One big estate like Nakambala in Mazabuka or Mpongwe on the copperbelt has a minimum of fifteen centre pivots as well as other overhead irrigation systems which are all driven by electric power. Those are massive infrastructure which need a lot of power to drive them to irrigate the fields and depending on the size of the field and the soil type, the crop being irrigated, they can run for about six to ten days nonstop to complete one cycle of irrigation. To drive water into such equipment, it needs pumps with great head levels to lift water and run it at high pressure. This is just one basic infrastructure that is needed in the production of the raw material like soybeans, sugarcane or wheat which are used in the agribusiness manufacturing industries. To supplement this piece of power supply infrastructure, the first republic put up another investment at Kafue Gorge which is another massive investment which is still under development to increase the capacity. This too, was supplemented with other smaller investments such as the Musonda Power Station on Luapula river at Mwense. However, that investment is not sufficient to even supply the whole of Luapula province, recently, it was being upgraded to increase its output. In short, the power generation, supply and distribution in the country is not sufficient. Although the country receives adequate units of sunlight, it has done less to tap into generating solar energy which can contribute enormously to the power deficit not only suffered in Zambia but in the region. Apart from South Africa and Mozambique, most of the countries in the region are deficit in power. Additionally, Zambia is a country with abundance deposits of coal but none of it was used in power generation, until late in 2014 when construction of a 300MW thermal generation plant was started and was anticipated to be completed by August 2016. This will hugely contribute to the power sufficiency and reduce on the country's power imports from Mozambique.

These power sources were adequate to supply the mines and the rest of Zambia then when the population was at three million with less industries and housing units. If there is anything that we slept on as a country and forgot about it for a very long time, it is the investment in power generation and distribution. The country boast of having the most water resource in the whole of the SADC region but it is unbelievable that most of the times it imports power from neighboring countries such as South Africa and Mozambique. We have failed to turn our comparative advantage we have in energy generation into a competitive edge. From the many days that have sunlight in a year, we have not taken advantage to develop the solar energy. None of our factories or machinery is run by solar power save for a few solar panels on the rooftops of a few houses which are used to warm water for domestic use in special domestic geysers. Countries like USA, India and unbelievably those in the parts of the world which receives a fraction of the amount of sunlight than we receive in Zambia, have developed this alternative green power source. Additionally, there is availability of the uranium deposits in the country but we cannot develop that because we do not have the technology and capacity to use such. As a country, we need to explore the many sources of power generation opportunities we have in almost all the ten provinces. We have the potential to supply the whole of southern Africa with power if we developed the many waterfalls that we have in the country. We have the potential to even earn clean green foreign currency than the one we are getting from the mines. Without adequate supply of power, there can't be any meaningful development in agribusiness and later on commerce. This came to the fore in 2015 when we witnessed the worst load-shedding ever. We were having not more than eight hours of power supply in towns like Lusaka and this is not very good for industrial and economic development. To emphasise the under development in energy, less than two percent of rural people have access to power. However, of late we have seen great strides by government to increase the generation capacity and trying to electrify the rural community though there is great opportunity for investments in this sector. We need massive investment in power generation and distribution if we are to see meaningful development in agribusiness and manufacturing industries in this country. There is political will from our leaders who control our resources but it is just that the pace is rather slow. There is ernomous business opportunities in power generation and supply because the demand goes beyond our borders. People that need to invest in Zambia must not look at energy as one of the growing sectors.

Communication and transport infrastructure – this can be divided into the road, rails, air, water and telecommunication infrastructure. Communication infrastructure like already highlighted was developed to some extent. Roads linking major towns especially those along the line of rail were developed. To state that the road infrastructure is well developed in this country would be like comparing to the proverbial saying of an ostrich burying its head in the sand and think it has hidden from being seeing. Agriculture just like any business needs good communication infrastructure. Giving an example of why the roads are important in agriculture; Lundazi and Chipata districts of Eastern province produces a lot of tomatoes. They have the potential to grow more tomatoes to feed the whole of Lusaka province, however, production of this very important horticultural crop has stagnated because of the distance to the main market in Lusaka and the Copperbelt. Farmers that have attempted to transport this commodity to Lusaka have ended up losing not less than twenty percent of their harvest during transportation. However, even if we have relatively good roads that link major towns, the feeder roads that lead into the high production areas are quite poor or underdeveloped. There is need to invest in feeder road development especially in high productive areas of the country. This is an opportunity that the private sector can take advantage of by partnering with government through public-private partnership (PPP) models.

While working in the Western province for intance, to cover a stretch of two hundred kilometres from Kaoma to Lukulu it used to take me not less than four hours in the rainy season because of the bad road. By the time I arrive in Lukulu, the first thing I would ever think of doing even before I eat was to take a nap, because I was very tired. These are the very important roads that will allow the smallholder farmers have access to markets. If traders are not able to access the productive areas such as Lambwe chomba in Chienge's district, then the crops such as rice produced in that part of the country will be uncompetitive or will lack market. Of late since this Patriotic Front (PF) government came into power in 2011, we have seen them continue to implement some of the plans left by the late president Mwanawasa's government and they have also put up new projects to open up rural Zambia. Zambia has since 2012 been a huge road construction site though the immediate economic benefits of some projects can be a discussion for another day. There are insinuations that they could have have been implemented for political expedience. We also need to be patriotic to our country by producing quality roads that can stand a test of time, at least fifteen years as opposed to some of the projects that are already falling apart. Foreign construction companies have got huge contracts with the government, though most of these companies are from Asia. There are so many opportunities in road, bridges and dam construction in this country to open it up.

Our rail line was one of the vibrant communication infrastructures we had before the late 1980s. I remember on several occasions using the trains to move from Lusaka to the copperbelt. We had two types of trains; the express one which only stopped at major stations such as Kabwe, Kapiri Mponshi and Ndola called Kafue while the other one stopped so often after every ten to fifteen kilometres, and it was called Luangwa or nicknamed as *Dobadoba* (literally meaning 'picking or stopping anyhow'). This piece of infrastructure was also used to transport heavy goods such as copper, fertilizer, cement and heavy equipment for the mines from one end of the country to the other as well as exporting to the outside world. To ensure that this communication infrastructure contributed to transporting agriculture commodities, side rails were constructed which led to major grain and input storage facility being built in such places such as the likes found at Natuseko in Kabwe, Ndola, Choma and many other towns along the line of rail. This helped to even sustain the lifespan of our roads. This is not the case today; our rail infrastructure is quite dilapidated and has become so inefficient due to years of neglect. The infrastructure has been neglected in as far as reinvesting is concerned. The country needs partners to revamp this important piece of communication. Probably what we may have lacked as a country is the policy on transport. For instance, we have seen people buying residential plots next to the rail line as trains are not regular on these rails. Many entrepreneurs have gone to buy trucks to transport heavy goods on the roads which have contributed to their dilapidation. Nonetheless, the current government borrowed some money to revamp the rail line though the infrastructure has never gone back to what it used to be in the 80s. The bulk of the resources borrowed to revamp this important infrastructure was used on other projects. What most of Zambians expected was that the government was going to procure some new trucks and wagons but they instead rehabilitated the old wagons that have outlived their usefulness. On the other front, we have also witnessed the exponential dilapidation of TAZARA which is run jointly with the Tanzanian government, running from Kapiri Mponshi of Zambia to Dar es Salaam in Tanzania built in 1975 by the Chinese. This was an alternative route to the sea port as opposed to going south.

This second rail line was built basically to be an alternative outlet to the sea; it was built from Kapiri Mponshi via Northern Province to the port of Dar es Salaam in Tanzania. This was built when the country had political problems with Zimbabwe which was called Southern Rhodesia then. This was a visionary thinking by our leaders then. This infrastructure has not spared in the years of neglect that followed later in the late 80s and early 90s when the economy of the country was on its knees. One would wish that we could have built another rail line to Western province linking it to Angola, although this could also have been neglected. I guess the reason the leadership didn't think of embarking on such a project was that it could have passed through Zambia's biggest game park – the Kafue National Park which has giant animals like the elephants, buffaloes and hippos including lions. This could have disturbed the environment for the animals besides having possibilities of fatal accidents. This could have affected tourism as it is the third most important sector in the country. The third rail line was built by the late Mwanawasa's government linking Zambia with Mozambique via Malawi. It was built to shorten the distance to the port of Beira in Mozambique. This rail line starts from Chipata. However, up to the time of this book's publication, this has been a white elephant as it has not been operationalized. Such as opportunities that can be taken up by the private sector to run as this can see a lot of decongesting of the port of Dar es salam for the goods coming to Zambia and the route may be shorter to get to Zambia. Of late the country is exporting a lot of cement to Malawi and this infrastructure can serve that purpose.

The third part of communication infrastructure which was relatively well developed was the tele-communication sector. We had a parastatal company called Post and Tele Communications of Zambia (PTC) which had the telecommunication and the postal infrastructure. This was before the advent of the computers and the internet. This was the only mode of communicating then, otherwise, one had to physically travel to relay the message. This this company, one would either send a letter, telegram or make a phone call. Post Offices were built in nearly all the districts and telephone lines connected to all towns and commercial centres in the country. However, today with the evolution of communication, we no longer used telegrams and very few people are using the post office for posting of letters. At the moment, people can write a letter and send it within a flash of a second to someone very far. Nearly all districts have mobile communication facilities although the quality of the service is what needs to be improved on by the service providers. The country have three international communication service providers with one local one. Encouraging to note that people are using these facilities to find out about markets in far flung places for instance. It is possible today for someone to know the prices of commodities in major markets like Chisokone in Kitwe, Soweto in Lusaka within a fraction of a second. This has eased the cost of doing business in the country especially in the agriculture sector. Internet communication is also one mode of communication which has eased the way people conduct business today. This though needs to be rolled out to outlaying areas especially the data communication. There are opportunities in this sector as we are already seeing agriculture extension workers using the internet platform to offer extension service delivery. Investments in innovative products and services that will support agriculture in the country are still available. The country has done so much to improve the communication infrastructure, however, there are so many opportunities for investment in the construction sector.

Storage and value adding facilities – storage facilities are very important in agribusiness development especially where the focus is value addition. Storage facilities are very important in agriculture as they reduce post-harvest losses caused by adverse effects of the weather and pests infestations. For instance, it would be very easy for someone to fumigate grain commodities in a silo than bags stacked on an open platform. To emphasize the importance of storage facilities, Zambia produced what was being called a 'bumper harvest' of maize in 2010/11 season. Nearly each and every district in the country had stacks and stacks on maize. All the silos and sheds were filled to the brim and the slabs in the open air were all stacked with maize such that the organization tasked with the responsibility of buying the maize grain had to improvise temporal slabs by stacking the maize on platforms made from tree logs. There was too much maize such that the firm ran short of tarpaulins to cover all the stacks of maize. Most of the maize was soaked during the subsequent rain season. It was estimated that out of the over three million tons bought, about twenty percent was soaked and condemned to be used either for human consumption or industrial use such as the brewing industry. I remember driving from Katete in the Eastern province to Petauke and just after Katete, there was a foul smell and decided to stop and investigate where the smell was coming from. To my surprise, I found a heap of soaked and rotten maize dumped in a quarry pit they used to mine laterite for repairing of roads. This was at the cost of the tax payer as a result of inadequate storage infrstructure. The heap contained not less than five metric tons of wasted maize grain. Fortunately, that year Zambia was the only country in the region to have recorded a surplus harvest of the maize grain which is the most sought after staple commodity in the region including many parts of East Africa. That same year, there was famine in Somalia but one wonders why we could not take advantage of the situation to sell the commodity to World Food Program or even donate it. Sad enough, this was a crop which was produced by smallholder farmers under subsidized inputs from the government and farmers got most of the financial resources used to subsidise its production as loans. On the other hand, many of the farmers that supplied that commodity to the grain marketing board called Food Reserve Agency (FRA) were not even paid on time as the agency was overwhelmed with supply.

This just shows that there is enormous opportunity in grain storage and value addition in this country. Nonetheless, of late the private sector has come on board in this area of agribusiness though at a snail's pace due to challenges of accessing financing. The last time the government had projects of constructing grain silos was before 1990. The few storage facilities that are being built now are warehouses with very minimal storage capacity and a few open slabs. The country is experiencing growth in the agricultural production especially for grain commodities like maize and soybeans which is being subsidized by the government. However, no equal investments is being channeled in storage facilities. Maintainance of such public infrastructure is also a big challenge as the government do no allocate enough resources to the ministry responsible. For instance, the silos at Natuseko in Kabwe were last used to store maize in 1993 and after that, there was nothing being stored because the policies of post 1990 did not favour agriculture. Some silos became breeding areas for bees and some snakes though this is not the case today. The country needs more storage facilities and private sector again can partner with government to manage such infrastructure or build more for commercial purposes. All this is an indication that the country does not have adequate storage facilities to safeguard its produce, and this is a great opportunity for those that want to invest in warehousing. For those who understand agriculture knows that storage facilities are basically the banks for farmers and they are very important for value addition too. This helps the farmers to secure their produce and only sell them when the prices are favorable for them to make better profits. Agricultural produce just like any other commodity follow the laws of supply and demand. When a lot of farmers are harvesting, there is too much produce on the market and this tends to depress the commodity prices and vice versa.

However, during the times of less, such as in the rainy season (starting in November), demand is more than supply and this pushes the prices up. This is the idea time a thrift farmer would want to sell their produce. For a farmer to keep his crop up to that time, it needs to be stored in a secure place and therefore, the storage facilities such as the silos are important in agribusiness. This is also important for governments because they have a function and responsibility to ensure adequate food supply for its population in the country. This responsibility is usually undertaken in market economies through governments putting in place the necessary infrastructure, and financial and legislative framework so that the private sector can operate. The markets on the other hand are required to be self-funding and to operate with a commercial approach for them to be sustainable. The success of any business depends on the reinvesting in it and this can only be attainable if that business is making profits. Additionally, it needs functional markets were the commodities can be sold from. This is what the subsequent government after the Kaunda government ignored to invest in. There were no policies formulated to create and enabling environment to allow investments into storage facilities for agricultural commodities. The country had been preaching about warehouse receipt system for a long time and this could not work if storage facilities are lacking. There has been great strides made to implement the warehouse receipt system in the country and the remaining jig saw to effectively implement it is mainly financing in order to have guaranteed markets for the commodities. This facility also presents a great business opportunity for innovative entrepreneurs to explore and partner with the institutions pioneering the facility in the country. This marketing and investment tool has worked so well in other countries in the region such as South Africa, Malawi and Kenya.

The other important attribute for storage as earlier alluded to is the need for value addition. To many people including some of our great agribusiness practitioners, they misunderstand value addition to mean processing only. Yes, they could be right to some extent but processing is just but one method of value addition. Processing is a type of value addition where there is total transformation of a commodity. However, value addition can mean sorting, grading and packaging. We can give an example here for us to be so clear with what we are talking about. I was visiting one community enterprise run by one of the cooperatives which grows and sell groundnuts in one farming community of the country. When I saw the quality of the groundnuts that they were selling, it had all sorts of groundnut kernels; broken, shriveled, rotten and so on in the same bag. They were looking for a market to sell the groundnuts more profitable than the prices being offered in the area. Upon being advised that if they wanted to fetch good prices for their groundnuts, they needed to add value. First they thought this was a far fetched dream for them because they did not have equipment to process the groundnuts. However, after explaining to them possible value addition methods, they were so elated. Indeed, they couldn't afford peanut butter making equipment for instance, but they could afford to sort their groundnuts to improve the market quality. An example given that if they sorted their groundnuts by removing the rotten, shriveled and bad looking nuts they could ask for a premium price for their produce. This could allow them even pay better dividends to the cooperative members. By sorting the commodity, they could even dictate prices to buyers as opposed to many scenarios where sellers (farmers) are price takers. Furthermore, the groundnuts could also be roasted after grading and sold as a snack as seen in many markets and bus stations in the country. Many of us think grading of such commodities or roasting in the case of groundnuts is not value addition. Many producers fear to sort and grade their produce because they think the quantities that will remain will be too little to make money but the opposite is true. The whole essence of doing business is to make profits and losses if we haven't done well. A business that does not make losses is not a real business at all. However, a good business shouldn't be making losses but a business always goes through cyclical graphs. The troughs in a business make the managers to wake up and strategise on how to improve on their operations. This was an indication that certain individuals in this business don't appreciate value.

I am one person who agrees with those that suggests that governments don't run businesses quite efficiently, however, they endeavor to make the business environment conducive for its citizens and the private sector to thrive. Normally, other countries have started big agribusiness corporations as a government and later privatized them. I think the government in the past had not done enough as a country to make the agriculture environment conducive for value addition. One extrinsic example I can give is the poor supply of electricity and unfavorable access to cheap financing in this country. I still do not understand, for instance, why there are two exchange rates for foreign currencies depending on whether one is selling or buying while other countries such as many in Europe use only one rate. There is huge opportunities in value addition especially of the agricultural commodities especially if the government can compel banks to support the private sector with affordable financing facilities as we have seen in other countries. Companies that would endeavor to invest in value addition will have a huge market opportunity because most of the goods sold on the market in this country, especially the processed food stuffs are imported from South Africa and other countries.

I will be down playing the strides the first republic did to make the value addition environment in agriculture conducive if I don't recognize the efforts of that government. In the first chapter of this book, I talked briefly about the efforts our government then made to establish at least one major industry in each province depending on the comparative advantages which that province enjoyed. However, these developmental programs were unfortunately not continued in subsequent governments that followed. We saw the haphazard privatization of many parastaltal firms that were formed by government. This has continued to this day and we do not have major agribusiness corporations that are owned by Zambians except for a few such as a few processing factories have been set up in the soybean value chain mainly in the livestock industry. We may discuss these enterprises in a bit more detail later but my main thrust here is to highlight the agribusiness opportunities that exist in this country in value addition especially in the manufacturing sector. Opportunities in this country are immense such that wherever one looks, opportunities are waiting to be harnessed and explored for investment.

One of the major constraints that has stagnated the agribusiness development in this country is lack of good and affordable financing opportunities. The few that have managed to put up value addition infrastructure which has survived to this day have either got financing from outside this country or had good connections with those that yield power in such institutions. Our banks' lending rates are just too high for businesses as earlier alluded to; they should be the highest in the region. One would think that the government did not have its own bank which could have been giving loans to its entrepreneurial citizens to establish value addition investment into the agriculture sector. The problem we have is that most banks that have dared set operations in this country only have retail products which are not conducive for this type of businesses. For instance, when the manufacturing sector was on the upswing we had a bank which was a wholly owned Zambian that was supporting such activities which unfortunately was privatized in 2006. This has left a void on the market. The banks in this country do not behave the way they do in other countries. Even those that have set base in the country, they have two types of lending rates; one for everyone and a lower rate for its citizens that have businesses in this country. This is the reason the local businesses cannot fairly compete with those owned or run by foreign counterparts. In the second republic, we had banks such as Export and Import Bank (EXIM), Lima Bank and others that provided relatively cheap financing to farmers. These had a specific purpose to serve; they were merchant banks with a view to growing the SME sector. However, some people might argue against this point and advance (work) culture of some of our citizens to have been very negative then and led to their downfall. This could have been true then but that has changed because of the current policies and political environment. The political environment then was socialism and the country had socialist policies which is not the case today. The bank that will dare develop products that are friendly to SMEs will grow or develop its business exponentially. At this important it is important to discuss how our culture had affected the business environment in this country especially as regards to agribusiness development in order to appreciate why we have the current agribusiness environment and how we can forge ahead to develop the sector.

Culture and work ethics – Tembo (2012:9)[1] in the MBA dissertation quoted Kreitner (2009)[2] as having defined culture as the collection of shared (stated or implied) beliefs, values, rituals, stories, myths, and specialized language that foster a feeling of community among organizational members. Organizational culture implies some of the

---

1 Tembo, F.H.M. (2012). *The impact of organizational culture on the team effectiveness in non-profit organisations: a case study of SNV Netherlands Development Organisation in Zambia*. MBA dissertation.

2 Kreitner, R. (2009). *Principles of Management*. 11th Edition. China: South-Western Cengage Learning.

traditions and values that are followed in an organisation and to an extent, this influences the way the members behave. It is worth noting that these important shared values and practices evolve over time and determine, to a large degree, what people perceive about their organization or country experiences and how they behave. This has been one of the biggest factors which have led Zambia to lag or stagnate in agribusiness development. Sometimes I sit down and imagine the day our mines will run out of copper, what will this country be? We have abundance of natural resources other than minerals which we could have explored and developed our industries to be competing with those of other countries in the region if not better, using the incomes from the mining sector. Well, we have seen how a mono economy can negatively affect the economic stagnation of a country in cases of price slumps. In 2015, the copper prices dropped to the lowest ever in a very long time of around $3,000 per ton coupled with inadequate power supply that the country was facing besides low global demand, production of copper declined. The resultant effect was a more than hundred percent depreciation of our currency against the major world currencies and great deterioration in the country's economy. This was because there was limited inflow of forex. Inflation increased from six percent in January 2015 to over twenty-one percent a year later. This had a negative effect on the quality of life of the Zambian citizens. Though partly this could be attributed to some political environment as the country had just lost its president and there were a lot of uncertainities by the business community. It can also be attributed to our biggest weakness which has been in the entrenched culture found in us of not worrying and planning for the future well in time. We tended to espouse a culture that do not want to grasp opportunities and are quite wasteful when we have more than enough. This culture emanated from the socialist cultures which were inculcated in the population when the country had much wealth than the people. It is sad that some people have refused to come out of this culture; *niva Boma* syndrome (simply meaning they public goods). They seem to have taken too long to note that with the advent of diminishing resources, the world was soon shifting to the capitalist type of ideologies; even those that were strong socialist adopted some level of capitalism where they blended their cultures to socio-capitalism. I have no doubt that even after the collapse of our economy in the late 1980s, had the we dared shift our ideologies and policies effectively, we couldn't be where we are today. Having a culture is another thing but what type of culture is it is another. If a nation has a strong positive culture, you expect to have positive results. However, the case of Zambia and its population is that a few of us have embraced a strong negative culture. Let me cite some examples to illustrate some of the negative cultures that the country has embraced.

You will agree with me that the culture of a company or a country will be mostly molded by the Chief Executive Officer or the the leaders of that institution. If the CEO of an organisation reports for work at half past eight in the morning, he or she should not expect the employees or the junior officers to report at seven hours. Nonetheless, the majority of the people are so committed and would want to see their company prosper, and they would put in a bit more by reporting early. Take for instance, the bank manager who always reports for work in a golf t-shirt, a new employee that joins the bank will think that it is official and soon everyone will be wearing anything in that bank. This was the case with Zambia. Unfortunately, this negative culture had been the case in our country for some time. During the first republic, we had a disciplinarian as the CEO of the country. The president never tolerated indiscipline from his ministers and government officers, let alone corruption. Many people remembers him for instilling discipline in his cabinet and the CEOs of parastatal companies. When making cabinet reshuffles, he is one of the few presidents that used to even mention the reason why he was dropping or firing a particular individual. However, over the years the country seem to have relaxed a bit and now we are seeing a new culture of materialism and corruption creeping into many Zambians. Being in vantage positions is synonymous with making money through getting contracts, corruption, pilfering and many negative vices. This culture slowly has crept into many institutions including the private sectors and it has become very strongly entrenched; this is the new culture that has been promoted for over twenty years of our multi-party democracy. Indeed, leaders form and shape a culture of an organization though the followers should also buy into it and not resist change. Nonetheless, it would be very unfair to the many Zambians that are upright and are trying so much to work hard by painting them as black when they mean well. It is a few people in vantage positions that have tarnished the image of our society by promoting negative cultures. There are still some very hard working Zambians that are promoting uprightness and all that they need is strong followers and leadership to espouse such values. They have made their money through hard work and they spend less than eight hours sleeping in a day. Good management is like a cylinder of oxygen to a patient. The leadership that was provided by the late Mwanawasa for instance, was to instil uprightiness and hard work. The country noticed glimpses of economic emancipation and orderly development. We only came to appreciate his hard work in agriculture besides other sectors after he was no more. This is not to say everything is lost, we have seen glimpses of a strong and awake civil society with good checks and balances, and all they need is support from institutions tasked with those responsibilities. With good leadership, this can change and the country will be back on the right track again. Corruption just like in USA and Europe is killing our development and it should be fought. The only difference is that we seem not to have very strong institutions to fight it in Africa but slowly the citizenry are waking up and standing to the challenge.

Education – this is another very important factor of economic development because it contributes to the skills of the workforce or human resource. Zambia is a country that adopted the British type of education system. We have the primary schools that starts from grade one to seven, thereafter, one proceeds to go to grade eight or form one as it used to be called up to grade twelve or form five at the secondary education level. Upon completing one's secondary school, they would proceed to go to college or university as it were. However, someone at this level is ready to do certain jobs such as being a clerk in a bank. What the country forgot to do in a long time was to review this education system because it was kept in its raw form from the way the British left it. Education just like anything else is dynamic, and it needs to suit the environment and culture of a particular society. There was less done to even revise our syllabus, for instance, and this has promoted the educating of people that are only ready to be taken up in white collar jobs or in public service. There was little done to develop the artisans or other skills.

Agriculture is a life science which needs a lot of research. In Zambia, we kept an agricultural education systems for a long time without incorporating new ideas such as the agribusiness and other practical skills development related to the primary products of the agricultural sector. For instance, the University of Zambia is the oldest and largest tertiary institution in the country. Up to as late as 2002, it did not have the agribusiness course as a specialization in that institution for instance. The school of agriculture in the institution only had four areas of specialization, which were; crop science, soil science, animal science and agricultural economics until 1999 when food science was introduced as a specialization. This is one limitation that has been with our education systems not only in agribusiness development but other specialisations as well. The other big challenge which could be the main reason not to bring on board other specialisations is inadequate and sometimes lack of funding to conduct research for such public institutions. The research department of any company is the lifeblood of that institution and if you are not keeping the research department active, you risk the products becoming obsolete as the case was with most of the research institutions in the country. For instance, Zambia was one of the pioneers in maize variety research and breeding in the region. At some point, one of the public seed company which was the component of the ministry of agriculture used to export seeds to countries in the region like Zimbabwe, Malawi and Zaire (Congo D.R). This was supplemented from the research by Zambia Agriculture Research Institutes (ZARI) under Mt. Makulu, Minsamfu, Kabwe, Mansa, Solwezi and many other such research places.

The country at some point had great scientists in government institutions some of whom later formed their own companies which are very vibrant. A good example is Maize Research Institute (MRI) which has got one of the best genome for maize such that it had attracted a multinational company which finally bought it in the fall of 2013.

This company was started by a maize breeder who was working for the government seed company. Nonetheless, it would be very unfair to ignore the hardworking teachers and lecturers of Zambia who under very hard conditions have been doing a very commendable job to educate the masses that are running the affairs of this country. The writer of this book is a proud product of such self-less hardwork. They have been trying their level best, although our standards have drastically been falling down since 1992. This speaks so well to some thinking of some great scholars that believe that people are not lazy but they tend to have impotent goals which do not inspire them. These could be true of these men and women of substance (teachers) that are highly demotivated because of several negative factors that are happening in our institutions of learning such as luck of teaching aids, low funding, poor infrastructure and generally poor work environment. Against these negative factors, they still have forged ahead to churn out students that are running the agricultural industry and the economy at large. In my view, the biggest failures in the education sector has been the support from relevant authorities; the government. For instance, since 2009 the country has experienced increased incidences of examination leakages in schools and some universities which has contributed to declining of the education system. This affects the quality of students that are produced.

However, it is worth noting that our education system is not all lost. It is still one of the best in the region and quite competitive globally. In the early stages of development, we had many neighbouring countries that depended on Zambians to provide human capital in their industries such as the education and mining sectors. Up to now, we still have well qualified doctors and nurses that are working in countries like South Africa, Botswana, Namibia and United Kingdom. Some of our finest mining engineers, agriculture extension officers and many other sectors are working in Angola, Namibia, South Africa, USA, Europe and Australia. Being a product of the Kaunda education system, I should mention that I benefited greatly from that education system. As a matter of fact, the education system of the first republic was so good that some syllabi have not been revised to date. People are not interchangeably as some leaders especially in companies sometimes think. Gifted people have unique talents and such people should not be forced into roles they are not suited for. This is evidence enough that our education system is still very credible but all we need to do is add a sphere of entrepreneurship so that we can graduate to be employers instead of being employees waiting for jobs from government. Additionally, the government in partnership with the private sector should effectively fund such public institutions. The responsible partners should play their roles by improving funding for operations to such institutions. Furthermore, a lot of resources need to be pumped in research and development if the services and products that the country is developing have to be competitive. There is still much more that need to be done in the education sector and opportunities for investment in this sector are massive.

The biggest worry currently with our education system though is that research has been neglected due to inadequate funding and the slow pace in developing and implementing new innovative programs. This was observed by the first republican president of Zambia[3] during the open day at the highest institution of learning in Zambia when he said, 'just like they tried in the times of 1960s, Zambia's universities should now be responsive to the environment of the 21st century. To be effective, universities must combine training and learning in theory with vocational and wholistic application. We must strengthen teaching, research and mentoring.' He further went on to observe that the university needed to be forward looking by embarking on programs that would respond to various local and global issues. This is true because the inadequate funding at the institution has made it difficult to effectively carry out programs especially those related to research. Most of the people that can afford have opted to send their children abroad to get education and I feel there are still opportunities in this sector for those that want to invest in Zambia. Other private universities and schools that have established base in this country are attracting

---

3 Mulenga, A. (2013). Be responsive to changing environment, KK urges universities. *The Post Newspaper.* August 17, 2013

many students including international ones from the region. This is a great opportunity for those renowned institutions of learning to invest in educational sector in this country to provide competition. One aspect that is also lacking in our education sector is the connect between the research departments of these institutions and the private sector. The products of research that are being developed at instiions need to be taken up by the private sector and upscaled or implemented.This is what we have seen in other developed countries like America, for instance, some of the posh vehicles we are seeing been driven were as a result of research by the military during the second world war and they have now been commercialized.

Politics and policies – to develop a country, there must be a good mix of technocrats, technicians and the political leaders. The latter have a great bearing on development as they are the ones that formulates policies which are implemented by the technocrats. In this country the actions of those in political power have a great bearing on the economic development. Politicians determine the destiny of the country especially in Africa where democracy seems to have a different definition to that of the western world. For instance, we have read in the print media and heard on electronic media what happened in Nigeria during the rule of one dictator where billions of dollar were externalized at the expense of the many people of that country. That country is rich in oil but the earnings that come from oil don't effectively benefit the local people especially the common men on the streets. This is not only perculiar to Nigeria. In our own country; we have not been spared of corruption. Although we have been one of the major producers of copper in Africa if not the largest for over two decades, very little benefits have trickled to the common man and poverty is ever increasing as production output of copper increases. There seem to be a correlation of the copper output with poverty levels in this country. Copper is a mineral that has been sought after for a very long time but we have not taken advantage of that to develop alternative sectors like tourism.

Firstly, in the first republic, our government tried their level best to develop our country using the earnings from copper. However, in the mid-70s, we went ofcourse and started using part of the earnings from copper to help liberate our neighbouring countries in the region who seem to have reciprocated less to our help. Some of these countries that benefited from Zambia's sacrifice to liberate them are Zimbabwe, Namibia, South Africa, Mozambique and many others. Our people lost their property and lives in trying to help liberate their neighbors. Zambia had paid dearly for most of the countries in the SADC region in the quest to help them get independence. Our infrastructure that are key in agriculture development and commerce were bombed and we had to rebuild them from scratch. Politics in Zambia has played a very critical role in agribusiness development. This is because the politicians literally control the resources of the country. Politicians are important in formulation of economic policies which affects the business environment of the country. For instance, in 2003 the government introduced consumptive subsidies in maize marketing where the government would set a farm gate price for maize at ZMW 65 per fifty-kilogram bag of maize for three seasons. Once the government buys off this maize from the smallholder farmers, it would offload the same product to the private millers at a reduced price of ZMW 60 per bag. The government's understanding was that it was trying was to bring down or control the price increases for mealie meal which is the product of maize processing so that it becomes affordable to many Zambians as it is the staple food. However, this act had been crowding-out the private sector involvement in maize marketing because they couldn't buy the commodity competitively. Many of the private sectors just waited for the government to mop up the commodity from outlaying areas of the country which were difficult and costly to reach. The private sector would then later buy it from the government. By so doing, the government would absorb the operational cost on their behalf. However, in 2013 the government decided to remove this subsidy but maintained the same price of K65 per bag. What happened was that the private sector did not wait to buy the commodity from the government. They felt the price was the right one and they went out there to buy from the farmers at higher prices than that which was set by the government before as the case was in 2016 marketing season.

In 1991 after the MMD government swept into power, removing Dr. Kaunda who had followed socialist policies for close to three decades of being in power, they did away with the latter's policies and introduced the capitalist policies. This introduction was so sudden and somehow, in regards to agriculture they were done with less regard for adverse effects. The impact was that the agricultural development in the country went down. This was not only because the country was going through a very difficult time economically, but the policies introduced of 'unbundling' anything that was done by the previous regime took the country backwards. One would expect that the new government would build on the foundation laid by the previous government but instead they replaced most of the things that they found. Certain policies that I feel they did very badly were to abandon the mid-term and long-term national planning. The implementation of projects was done haphazardly. The results were that our agricultural productivity and production of all the crops went down and we lost certain great markets such as the European Union (EU) for groundnuts because the quality and quantity of the produce was compromised.

After the late Dr. Chiluba's ten-year reign with his *nu-kacha* (new culture) policies, the late president Mwanawasa took over the mantle of leadership. I have a feeling that Mwanawasa did far much better in his seven years of being president than what his predecessor did because he espoused the attitude of development based on the principle of building on while innovating. His government was based on the principle of rule of law and stopping corruption, which is a big nuisance to most of the governments not only in Africa but world over. I wouldn't be wrong to even think that he would have even done much better had he been the one that had taken over from the first president. The reason being that the way he was ushered into office in 2001 was questionable and he had a limit to which he could fight corruption without exposing the systems that were used to put him in office. This is like using a baby blanket to cover your body in the cold season, as much as you would stretch it there are certain body parts that would not be covered because it is small. Nonetheless, the man tried his level best which was good enough in the opinion of many Zambians. Most Zambians saluted him and his team to have re-introduced mid and long term planning at the ministry of finance. He developed the fourth and fifth national development plans in addition to the vision 2030 document. His strides to plan ahead helped so much in that this saw the planning and implementation of so many infrastructural developments very important to economic emancipation and the country at large. This is not to insinuate that his predecessor did not do anything but the fact that his administration was mulled in corruption towards the end of his reign subtracted a lot from his leadership.

In agriculture, Mwanawasa introduced many policy documents such as the national agricultural policy (NAP) of 2005, the fertiliser input support program (FISP) which saw Zambia move from a grain deficit to a surplus and the vision 2030 which is being implemented to date as a long term plan. Some of the great policies that helped boost the markets for the local produce such as soybean value chain was the instituting of a ban on the exporting soybean as a grain. This policy saw investment and the ultimate expansion of the soybean crushing capacity for the country from a partry 120,000 metric tons in 2007 to over 300,000 metric tons by 2011. This policy encouraged the private sector that were involved in exporting soybean grain to put up crushing facility as they were only allowed to export the cake and cooking oil as finished products. A great number of processing companies in the livestock feed industry came on board, some of which are; ZamBeef through its Zamanita plant which increased the crushing capacity by installing new equipment, establishment of factories by new entrants such as Mt. Meru in Lusaka, Emman in Luanshya and Gourock Industries of Ndola on the Copperbelt. This increased crushing capacity and moved Zambia from a net importer of soy grain to a net exporter of the cake as a by-product of oil extraction. Many farmers started producing soybean because the market was readily available. A crop that initially used to be grown by commercial farmers only is now being grown by smallholder farmers as well.

Most technocrats including our political leaders have been championing diversifying of crop production from the traditional maize to other cash crops such as soybeans, groundnuts, sunflower, rice, tobacco and cotton, even into rearing of livestock and fish farming. I should state that for certain reasons I don't understand quite well, why we are good at formulating policies yet very bad or slow at implementing them. For instance, we started talking as a country about diversifying around 1990s but little or no enabling environment was created to allow for the real diversification. It was only until 2015/16 production season that we saw the real diversified support from government to small scale fertiliser input support program since its inception in 2002. Before then, the support of the subsidized inputs had been centred on mainly maize production. This to some extent has led to failure or slow pulse of adoption of other agricultural support interventions such as the crop rotation amongst the farmers. Ultimately, this has led to degradation of some soils that has been highly mined of certain soil nutrients leading to stagnation in productivity improvement for the country.

Natural resources endowment – Zambia is a country that you could compare to a 'child born with a silver spoon in its hand'. The country is endowed with so much natural resources; anything that is needed to prosper in agriculture, mining power generation, tourism, manufacturing, service industy, name them; the country has. The country has land which has the potential to produce any crop grown in the sub-tropics and support animal life. The soils are fertile ranging from the famous molisols to the oxisols as well as the marginal lands that are suitable for ranching only. The country has abundant fresh water resource for irrigation and transportation. There are four main water bodies or lakes with fresh water besides the so many lagoons as earlier elaborated. Additionally, there are many rivers and streams in almost every province of the country. The weather is favorable with plenty of sunshine to support many different types of crops. The country has several dambos with sweet grass that is good for supporting grazing and watering of cattle. Additionally, the country has a youthful population eager and energetic to work. There are other supportive industries such as the mines that need a lot of agricultural produce besides them bringing in foreign currency that can be invested into agriculture. There are eighteen national parks and game management areas well stocked with wild life including the big five (elephants, hippos, lions, rhinos and giraffes). The reasons why this potential hasn't been harnessed to the fullest could only be described as a myth or is it a curse? The forests are rich in hardwood that can easily be used to develop the timber industry. Some of the finest rosewood you would find in the world are coming from Zambia as well the as the famous mukula tree which is being explored for the Chinese market. Not long ago it was discovered that the country had one of the best timber which is being exported to China to make furniture supplied everywhere. Some of the gun butts and interior of posh cars are made with timber coming from Zambia. The country has over five waterfalls that have the potential to be developed into hydroelectric power stations for generation of electricity. There is abundance amount of sunlight through out the year and little is used for generation of solar power. The country has people that are so friendly and welcoming such that most neighbouring countries that have conflicts run to Zambia as refugees. To some extent the political leadership has not been forthcoming thereby letting down the country as far as development is concerned. It is unbelievable that in a land of abundance as demonstrated above, there are found some of the poorest people on the continent; people that can barely survive. Now that the world is contemplating commercializing the making of electric cars, Zambia has got Zinc, Cobalt, Manganese and copper which is key in this industry. There are several other minerals in the country such as lead, uranium, gold, diamonds and emeralds that of the best quality one can find anywhere. In short, Zambia is a paradise on earth.

Inherent production technologies – Our forefathers were the pioneer of some production technologies on crops practiced to date. Many people would think application of fertilisers in crops was introduced by the foreigners as they came to this part of the world in search of land and minerals. However, our forefathers and parents had used sustainable technologies in production of their crops even before the whites commercialized agriculture in Zambia. Some of the technologies they used were so sustainable that modern science is slowly going back to use them for research. The only limitation was that they were not able to explain the technology and reasons why they were doing certain activities. For instance, to some extent they knew that intercropping agro-forestry plant such as *Tephrosia* and *Sesbania* helped to improve the soil fertility, but the reason it was intercropped was not for the sole purpose of soil fertilization. This plant, *Tephrosia* was natured and grown amongst the crops for the sole purpose of providing a poison which they used to kill the fish during the dry season. The leaves from this plant would be pounded and 'washed' into water in a pond that had fish. Within a few minutes of making the solution, fish will be killed and it will start to float to the surface to take in the fresh air and they would scope the fish with special baskets made from reeds and grass. The small fish would die first while the larger ones like the breams will follow later probably after about two to three hours. The strongest ones would be the barbel fish which would die after four to five hours. However, it has later been learnt through modern science that *Tephrosia* is an agroforestry plant which has nitrogen fixing properties. As a young boy I remember so well how grandmother would treat me if I destroy this plant. When I become tired and wanted my grandmother to relieve me from weeding in the field so that I can go and play my favourite game - football; I would accidentally chop off a *Tephrosia* plant as I was weeding the sorghum. Sometimes, that deliberate 'accident' would land me a nasty beating especially if I chop off a well grown plant. This was not the case if I did the same to the sorghum plant; I would just be cautioned to be careful. This to me explained that she valued the *Tephrosia* more than sorghum. Our forefathers were indeed the initiators of some productive technologies for conservation agricultural techniques which the white man has refined. They seem to know what they were doing but could not only explain them.

Market opportunities – a market can mean a place for selling and buying goods and services, while on the other hand, it can mean the systems and infrastructures whereby players engage in exchange of goods and services. While parties may exchange goods and services through barter, most markets rely on people offering goods and services in exchange for money from people that need such goods or services. It can be said that a market is the process by which the prices of goods and services are established. For productivity and production to be improved and increased, there need to be a market for any product or service being offered. It is a known fact that the presence of market opportunities, 'pulls' production of goods and services for any industry. This has happened on several occassions in our country as regards crop production. The country has promoted production and consumption of maize for a very long time and this has come at the expense of other cash crops. There has been an increase in maize production because of the corresponding increase in the amount of land put to maize production. This has come about because of improved marketing systems for the commodity though productivity has stagnated. The government has created a conducive environment to provide for the maize market. Zambia's climate and the soils can support production of so many crops and rearing of different types of livestock. Crops such as maize, groundnuts, sunflower, soybeans, horticultural crops, bananas, citrus fruits, cotton, cashew nuts, sorghum, millet, rice, sugarcane, pineapples, poultry, macadamia and livestock such as cattle, sheep, goats and some aquaculture have been produced in the country. Additionally, many more agricultural and forest products can easily be harnessed in this country. To date, the major commodities that have been commercialized are maize, wheat, poultry production, a bit of cattle, groundnuts, cotton and of late soybeans and sunflower. The country does not only have the potential to produce the said commodities but it also has the best comparative advantage in the region. One may wonder as to why the country has not fully explored the comparative advantage it enjoys. Besides so many other reasons, one of them is lack of sustainable markets or too much reliance on government to provide the market opportunities? To some extent, one can insinuate that the country lacks the entrepreneurial mindset.

Remember that we have highlighted a lot of factors that favors the development of agriculture in Zambia and some of the factors that work against our country are the low population and generally poor infrastructure especially the roads in rural areas which luckily, the government of Zambia is addressing. For a market to function well there should be demand and well developed infrastructure that supports market dynamics. A good example to understand how important the market is in agriculture development is the maize value chain. For instance, in 2010/11 production season the area under maize production by small-scale farmers who are in the majority and produce much of the maize, sunflower, sorghum and other crops consumed in Zambia with an exception of wheat which is predominantly grown by commercial farmers, planted 1,355,764 hectares under maize. With a national average yield (small scale) of 2.0Mt/ha, the estimates gave a potential production of 2,711,528Mt. However, the national production output that year was over 3,020, 380 tons. In our country, value addition as far as maize is concerned has been well developed. The millers grind or process the commodity into mealie-meal and some make livestock feed, and part of the commodity is used in the brewing industry to make some opaque beer called *Chibuku*. The disturbing piece of information that a business minded person wouldn't want to hear is that all this maize had to be bought by the government that doesn't even own a single milling plant neither do they own a chibuku brewing industry. This ultimately affects the value addition activities as the country has continued to see the government selling the same maize to countries in the region as a raw commodity. The country is a member of the COMESA and SADC groups of countries. These two economic regional bodies have a population of 390million and 280million respectively. The member countries have signed trade treaties which makes the movement of goods across member countries seamless though this needs to be developed further. Zambia therefore, needs to explore this huge market potential for its agricultural products. The products from Zambia have opportunities to be sold in both COMESA and SADC countries without major trade restrictions. This indeed, is a yawning opportunity for the private sector to invest in. For instance, the cornflakes that are consumed in the country are all being imported from South Africa because of the limited value addition base. There are huge opportunities in the country for value addition to agricultural products.

The government has been buying agricultural commodities especially maize and rice through an established marketing institution called Food Reserve Agency (FRA). This was established through an act of parliament as a buyer of strategic reserves for the country. This body establishes satellite depots in almost all the districts in the country. By so doing, they create demand for the commodity at farm level and this has led to the country producing more than it needs since 2003. However, with the amount of land put to maize production, the country can produce more if productivity is improved. Additionally, when the world market price for the commodity was just under US$200 per metric ton (US$190 in 2012), the Zambian government was offering at the price of US$260 per metric ton at farm gate to the Zambian producers. This encouraged the farmers to produce the commodity as it was more profitable. Even for those farmers that were found in areas where maize could not do well inherently, they were encouraged to grow it due to the price factor. In commodity trading especially agricultural related, there are four factors that are considered and these are; price, volume, quality and consistency. Since the farmers produce inefficiently, which affects the quality and quantity traded in, the government has helped them to remain in business by twinkling with the price. Additionally, to further encourage the farmers to grow more maize, the government has also been providing production subsidies by selling fertilizers at K50 per fifty kilogram bag instead of the market price of between K180 to K250 per bag (50 kg) as of 2012. There is a reason the government has been offering such unbelievably higher than world prices for this commodity. Most of the small-scale farmers produce maize inefficiently because their yields do not meet the gross margin for the commodity; meaning their production costs are higher than what they can get at ruling world market prices. This is because their productivity is far too low; to break even with high input prices in Zambia if bought on the commercial market. To break even, one need not produce below three to five metric tons per hectare. The subsidies from the government lowers the input costs thereby allowing the small-scale farmers make a marginal profit for them to continue in production. We know that such policies are done even in developed countries like the USA but they are done so smartly than we probably do it here. They call it smart subsidies. The largest opportunity to grow the market lies in improving productivity because then, the farmers will be able to competitively compete with efficient producers from developed countries for the African market as well as that in China.

Comparing the above statistics to that of soybeans in the same marketing period; the country cultivated a 61, 422 hectares of soybean. Over eighty percent of this crop was planted by commercial farmers. The total production of the commodity was 116,539 metric tons. The total national demand for this commodity was 300,000Mt against maize of 500,000Mt. However, the quantity of maize produced over what was demanded was more by eighty-three percent while soybeans was under produced by sixty-one percent, meaning there was still excess demand for soybean unlike maize. The question that might seek answers is why the farmers did venture into production of maize even though the demand was limited? There are so many reasons for this skewed production towards maize and some of them have been stated earlier in this chapter.

To sum up, one reason is that although both commodities had the available market, the former normally has a better price especially if bought by the government and it receives so much support as compared to soybean production. Therefore, it is true to say attractive markets generally stir up production. In our country, most commodity markets haven't been that attractive especially those found in the rural areas due to the poor supportive infrastructure such as the roads, storage facilities, market information systems, extension services and so on. The heavy involvement of the government has made the market unattractive to the private sector. Additionally, it should be mentioned that the market is unsustainable and this tends to crowd out the private sector who are supposed to be drivers of the process. Nonetheless, we have seen in the some marketing seasons spikes of perfect market systems such as in 2016 when there was less involvement of government in marketing of maize. Though they set the FRA buying price for the commodity, they went into the market late to allow the private sector participation; normally this is how markets are supposed to work. However, there was still a lot of influence; much more has to be done such as restricting the government from being the only main entity involved in exportation of maize to outside markets but open it up to other private traders as well. The private exporters could not compete with government because despite buying the maize at a higher cost, they offloaded it to those governments at lower prices. The government also puts up so many export restrictions which discourages the private sector. These types of inconsistencies in policies and business hurdles have contributed to agricultural stagnation in Zambia with very little involvement by real agribusiness entrepreneurs in the maize sub sector. However, on the other hand it should be noted that most of the goods that are consumed in Zambia with an exception of a few are imported. For instance, we import all the clothes that traded in the country. The country has built some shopping malls that are stocked with mostly imported goods. This is a great opportunity for the Zambia industry as we need to substitute the market that is being taken up by goods from other countries with Zambian products. The real opportunity lies in adding value to the commodities that are churned out of the agricultural sector so that they can displace the imported finished goods.

Socio-cultural environment – In sub-Saharan Africa, over eighty percent of the food is produced by smallholder farmers on family owned farms. The inherent production systems are predominantly subsistence with less advanced technologies used. These farms are less mechanized and use mostly family labour for production. They is need to mechanise production and adopt certain technologies to improve the efficiencies. The labour supply has been affected by the impact of HIV/Aids as people are suffering from the disease. There is also high predominance of malaria especially amongst mothers and the children who provide most labour on the farms. The productivity per farm is very low, with most crops recording less than thirty percent of potential yields. In the last twenty years of agriculture in the region, we have seen a few countries steadily improving their agriculture industry but a lot need to be done. The bulk of the population falls in the age group fifteen to sixty years. The continent still remains the next frontier in agricultural development due to the abundance of land and other productive resources. Its population is inherenetly young. Nonetheless, it would be prudent to understand how agriculture was practiced in the country before independence in order to understand its evolution or development to date.

Agriculture in the first republic

Upon attainment of independence in 1964, the UNIP government and its leadership put up a campaign policy to nationalize the major industries in the country that included the mining companies. They wanted these companies to be in the hands of the locals as way of owning the economy and empowering the population. Before independence, the mines were in private hands and run by the whites. Much of the resources were externalized because the headquarters of such companies were in foreign countries. Some of the companies sub offices were based in Zimbabwe and South Africa. There was some unsubstantiated thought that the white government got labour from Malawi to work in the mines in Zambia and the resources were sent to Zimbabwe to develop that country where the administrators were based. This explains to some extent the reason why that country has better infrastructure than Zambia besides other reasons which borders on leadership.

Precisely, it is not very clear when maize was introduced in Zambia, but my great grandparents told me that it came with the Portuguese traders through Mozambique. Zambia was used as a transit for goods when trading with the east. Mind you before that, our parents relied on sorghum and millet as the main source of starch. Protein wasn't a problem as wild life and fish were readily available in the wild and rivers. There were no restrictions on hunting as there were more wild animals than people then. In some parts of Zambia, they used to grow cassava and sorghum even as late as 1980s as the only starchy crops and some have continued to date. For instance, my grandparents completely switched from growing sorghum in 1986, before that the main crop in the field would be sorghum intercropped with some maize which was mainly eaten whilst green and some of it would be used for brewing of local the beer famously known as seven days (*katata*). Sorghum consumption apart from the health benefits due to the special carbohydrate content it has and some glutein, it is also very easily to grow and it quite tolerant to droughty conditions. I guess the pull for my grandparents to completely do away with growing of sorghum was due to the market opportunities created by the National Marketing

Board (NAMBoard) which was buying maize. Maize growing received a lot of support from the government in form of input loans, especially to the smallholder farmers. However, even though sorghum was not bought by NAMBoard (not to my knowledge) it had a ready market from the brewers of the local beer as well as the breweries and it was the staple food too. My grandparents would despise a meal (*nshima* - thick porriedge) made from maize that it became cold quite fast and it never used to keep long in the stomach. Actually my grandmother would blend sorghum and maize when going to the grinding mill in the ratio of two to one. I guess this was the initial introduction and commercialization of maize. Even though I was young to drink any alcoholic beverage then, I guess beer made from sorghum was too strong than that from maize as could be seen by the way people got drunk. The other reason that could have pushed Zambians to move from production of sorghum to maize was the campaign mounted by the private sector. The only institution that was supporting research in the breeding and agronomy of sorghum was the public sector through Zambia Agriculture Research Institution (ZARI) at Mt. Makulu. However, on the other hand, we had a lot of companies in the private sector that came on board to support the research and development in maize breeding. Furthermore, our own government promoted the research of maize more than any other crops in the sector. They started giving loans in maize production and offering ready markets for the commodity. Then the country had institutions such as Lima Bank which was solely created to give loans to the farming community especially the resource poor Zambians but viable smallholders. These all activities points to the strides to commercialise maize which were successful. I believe that if the same formula can be used on cassava, the results will be positive too. It is good to note that cassava though a heavy feeder, does not require so much inputs as it is an efficiency crop.

However, during the government that took over from Dr. Kaunda led by the late president Dr. Frederick Titus Jacob Chiluba popularly known as FTJ, the country took a complete paradigm shift in the way the economy was being run especially in the agriculture sector. There was emphasis to put all productive resources and companies in private hands. Surprisingly, this was extended to even those that were doing fine. There was a complete shift to privatization even though some entities only needed to be recapitalised and restructured as opposed to complete sale. The unfortunate part was that some of these institutions were sold at values less than their economic values and many of them were either relocated to other countries or had their equipment and machinery ripped off and sold as scrap metal. This act made the Zambian farmer to lose the market opportunities because agro-commodities were literally sold as raw materials to outside industries without any value addition. This did not go well with most Zambians and the subsequent governments as corruption was perceived to be at the centre stage. The country had chain stores like Zambia Consumer Buying Corporation (ZCBC) and Mwaiseni stores which were stocking Zambian products processed within the country. Companies such as Dairy Produce Body (DPB), Refined Oil Products (ROP) and many others which are now no more are the ones that were adding value to the commodities generated from agriculture sector. Besides providing market opportunities, these entities provided employment opportunities to many Zambians too.

Many people agree to the suggestion that privatization could have been done slightly differently and hopefully in phases so as not to distort the available markets at the time. These institutions would have been giving competition today to the South African chain stores that have taken over the markets in Zambia such as Shoprite and of late those from Botswana called Choppies. These chain stores have taken most of the retail business in Zambia and indigenous Zambian businesses have been left to be street vending because they cannot compete. Some of the buildings for ZCBC for instance, were bought off by Shoprite chain stores that is mainly retailing imported products from South Africa. By so doing, they have been depriving the Zambian producers of the most needed market opportunities. This means commodities like sunflower which was the main raw material in the manufacturing of cooking oil by companies such as ROP had no market and many farmers stopped growing it. The country started importing such products and others from the countries in the Far East as well as East Africa besides South Africa. The government emphasis was left on maize production because this is the staple food crop and it was one commodity which sparked food riots for instance, that led to the first president losing power in 1991. Though we are seeing a bit of development in the agriculture sector by the local people as well as the manufacturing sector but they are unable to effectively compete with cheap imports from foreign countries. Many reasons have been advanced for this but the main one is the high cost of doing business in Zambia. The main constraints for the locals to attain the competitive edge has been the high cost of acquiring finances from banks to invest in various businesses. Many of the agripreneurs that have endeavoured to take the risk through borrowing at commercial lending rates from banks have ended up loosing their properties and farms to the banks. In this book we will endeavour to highlight this challenge of lack of access to cheap or affordable finances in the later chapters. Access to finance is one key element for effective agribusiness development and if it is lacking in the sector, no meaningful development can take place. On the other hand, other factors such as market opportunities are equally important in agriculture because everything that is produced must ultimately be sold if such businesses have to be going concerns.

Agro market development

It would be very unfair to the government of the first republic to talk about the few mistakes they committed at the expense of the many good works they did to promote agricultural development in Zambia. Former president Kaunda and his team notwithstanding the moderate education background they possessed, they had a great vision for this country. After independence, many of the big commercial farmers were white settlers that had bought big chunks of land along the line of rail. It was not by mistake that these settlers had settled along this stretch of land in the country. This was not only the fertile land but it was also strategically located close to the main transportation corridors; the rail line. One may wish to know that the biggest market then for agricultural outputs was on the Copperbelt where most of the people had trekked to go and work in the newly opened mines. Agriculture being an industry that needs good transport infrastructure especially if dealing in perishable products with shorter shelf life such as vegetables, it needs an efficient transport system to the market. This was one reason the white settlers first settled along the line of rail even in marginal lands at the expense of the fertile lands on the periphery such as that in Kabompo, Mpongwe, Shangombo, Milenge, Lundazi and many other regions in the country.

To evenly distribute development, the leadership came up with a master plan to decongest the line of rail and to ensure each and every part of the country received equal share of development. By developing all the parts of the country, they were preventing people from moving from one part of the country to go and seek employment on the Copperbelt for instance. This was a strategy to also prevent urbanization and to try and equally share the national cake. They must have known that the only two industries that were moderately developed then in comparison to others were the mining and agricultural related industries. They knew that one day, the minerals will run out and if they do not develop other alternative industries like agriculture which is renewable; the country risked reaching a 'dead end'. The government coined a very important terminology which they were calling going *back to the land*. They looked for comparative advantages of each region and developed strategies to create a competitive advantage for different agro-industries. For instance, Mwinilunga had favourable conditions for production of pineapples and there could have been people already growing the crop on a subsitence level for food and a bit to sell. They government therefore, created a pineapple canning factory in Mwinilunga. This already was a market created for all the people around that area who were growing and were going to grow pineapples. This encouraged more farmers to grow the crop as well as others to join in the value chain. The economy of the district was centred around this industry. It made more economic sense to have pineapples processed in Mwinilunga than having them transported all the way to Copperbelt in the raw form. Transporting the produce to the Copperbelt could lead to more post harvest losses and inefficiencies in the industry.

In Western province, the climatic conditions and the environment is favourable for plantation crops besides others such as rice production. The government established a cashew nut industry in that area. It economical to grow this crop as the place is relatively drier with poor Kalahari sandy soils that could not profitably support growing other cash crops like wheat and soybeans for instance. A cashew nut plantation was established in Mongu on an out-grower scheme basis providing more opportunities for the locals. Additionally, with the help of the Japanese, a canal to be used in rice production was established at Sefula along the Zambezi river banks which was irrigating the rice fields. Though the project did not receive the desired support by the government as it was initiated and started by support from JICA. It has remained underdeveloped and hasn't benefited the locals as it could have. Had it been well developed Sefula would have been a district by now and different crops such as vegetables and citrus trees could be growing there by using that irrigation infrastructure. However, the rice production has continued in that part of the country and Mongu contributes to about forty percent of the rice grown in the country. There is still enough potential in this value chain to produce more for the country and for the export market. The rice belt for that region has expanded from Yambezhi district near the source of the river Zambezi to Sesheke district in the south bordering Namibia. The eating habits in the country are changing and more people are turning to rice. Therefore, there is more potential for this crop. The country does not produce enough of the commodity and it imports the shortfall from Thailand, India and other countries.

In Southern province, there was potential seen in the Kafue plains for plantation crops such as sugarcane. Nakambala sugar factory was mooted and established and so many out-grower arrangements were established around that investment to provide market opportunities for the cluster of farmers. This plantation stands to this day and it is one of the major industries in Zambia contributing thousands of jobs to the country. It also supplies over ninty percent of sugar consumed in the country and exports some to the Great Lakes region and Europe. To support this industry, tertiary industries were developed. These industries created a market demand for sugar besides export. Though the company has now been privatized, it is contributing so much money to the treasury in form of taxes to build this country not to forget the employment opportunities the investment has continued to create for the Zambians. The small town of Mazabuka wouldn't be what it is today without the sugar industry. The town is south of Kafue; another town in which manufacturing investments was made. The government established a secondary agricultural factory which was manufacturing inputs to support production of the agro-industries that they had established all over the country. Nitrogen Chemicals of Zambia (NCZ) was built to be making basal and top dressing fertilizers to supply the agricultural industry in addition to the explosives that are used in the mines. They were planning using the value chain model to ensure that they support the entire activities in the value chain.

This was the case with the cotton sub-sector where they supported activities starting from the crop production itself and made backward linkages as well as vertical integration and forward linkages too. They could have thought that producing cotton for export was not good enough at all but to add value locally. Creation of tertiary industries was a way of keeping wealth and to grow the economy within the country as well as creating jobs for the ever-growing population. With this in mind, Mulungushi Textiles in Kabwe was created that was buying the lint cotton and processing it. Processing cotton into linen was not enough; the country needed to add more value and Kafue textiles and the one which was based in Livingstone were established. The livingstone textiles used to make blankets while Kafue and Mulungushi were well known for Chitenge and clothing material. No one would dared to wear the wax chitenge from Zaire then as the case is today because our products were quality and superior. Other subsidiary industries like Serios, a company which had specialized in making suits and Lusaka Clothing Factory which was making uniforms and other industrial garments were formed too. These industries were meant to be buying semi-finished products from the secondary industries. In Luapula, the government understood that bananas can do well in that area and they saw no reason why Zambia could be importing bananas from countries such as South

Africa and Mozambique, let alone Uganda and Tanzania. Of late we have seen bananas coming from as far as Costa Rica in South America. This is absurd because the country has a perfect climate and soils that can support banana productin. The government established a banana plantation at the border of Mwense with Kawambwa districts called Munushi Banana Scheme which was supplying many factories and supermarkets in the country. This came with a full processing factory making such products as jam and juicy which were of international standards. Kawambwa district also had its own industries; the first being the Kawambwa Tea company which had a plantation and a factory that was processing tea into the finished product ready to be exported. The tea brand was called Tanganda and foreigners used to like it very much because of its aroma. In addition, even though the people of that area are not good herdsmen, the government gradually introduced livestock to the area by establishing a ranch at Chishinga in Kawambwa and Mbesuma in the Northern Province. The province has great potential for livestock because of the sweet grass, abundance of water and it is sparsely populated. Going up further to Nchelenge, are found people who are called in the local language as *baTubulu* literally meaning people who sleep on the lake fishing because this was their most activity then though they would in addition cultivated small fields for cassava to provide starch. These are fishermen and the government established a fishing plant which was to extend the shelf life of the fish products at Nchelenge called *Tambabashila*. It used to supply fresh fish to the entire country. This company was later privatized and sold to the famous former governor of Katanga in Democratic Republic of Congo.

However, even though I had earlier mentioned that the government had concentrated on promoting maize subsector more than other commodities, this was done towards the end of Dr Kaunda's term. In marketing there is a strategy to market to your best customers first, your best prospects second and the rest of the world last. May be the government had a vision that with time, feeding the world would be a problem, so it needed to perfect the art of producing maize so that later in the years, Zambia would be a net exporter of food and corn in particular. For instance, sunflower was a crop that was promoted and grown by commercial farmers from Kalomo to Kabwe. This was because the crop had already established market at Refined Oil Products (ROP) where they used to process it into cooking oil. Had this company not been sold through the privatization which is still a topical issue today, the sunflower value chain would have been as well developed as the soybean sector if not more. During that time, soybean production in Zambia was just been introduced and it was grown by a few commercial farmers. People didn't like the oil from soybeans as some claimed that it had a certain pungent smell but today, it is the main source of oil as well as the cake for the feed industry.

The first president did not only think of the agriculture industry, he also thought about the other industries to do with commerce. In Mansa of Luapula province, deposits of manganese were discovered and this was a key raw material for production of batteries. The government established a factory which was mining manganese and making batteries branded as Spark Batteries! This industry besides others employed a lot of people from the communities around. When you go to the east, the government created industries as well. Due to the bad terrain and probably the poor road connectivity in that province, they established a bicycle assembling plant called Luangwa Industries. The decentralization though partially done was vivid with establishment of industries in all areas. For instance, they knew that Zambia Railways was a big industry; they decided to establish its headquarters in Kabwe to serve as a transport facility for the first mine to be opened in Zambia in 1921 as well as support the economy of the area which only depended on agriculture.

In a nutshell, post-independence agriculture in the first republic was done with well thought after vision by first government creating markets for the various commodities. It was well thought after and judiciously followed a value chain approach. Nonetheless, towards the last ten years of Kaunda's rule, he had fallen out of favor with the western world. For those that were old then, they claim that things started going wrong as early as the 1973 or somewhere about there. This could have been due to the pan-African ideas he had embraced of wanting to help liberate the region from dependence on the western world and colonialism. He waged diplomatic wars to help liberate countries in the region such as Zimbabwe, Namibia, Botswana, Mozambique and South Africa that were still under the white colonial governments. However, even if things started going sour then, the worst impact were felt sometime around the 1986 when riots and strikes were the order of the day. This was characterized by printing of money as the IMF and World Bank squeezed the government by suspending bilateral and multilateral support. Inflation was the order of the day and shortages of basic commodities such as soap, sugar, salt, margarine as well as cooking oil. Though the industries were still manufacturing these, they were being smuggled to neighbouring countries like Zaire and others. During the period 1978 to 1990, there were about three to four unsuccessful coup attempts on the government. The latest which I vividly remember was by a junior officer in 1990. It was evident that the government of Kaunda was numbered and coming to an end so fast. Agricultural development was not spared and that led to shortages of mealie meal, the staple food crop for Zambians which was compounded by smuggling of the commodity. One would not understand as how we could have had those shortages when a year earlier, in 1988/89 season, Zambia produced the biggest bumper harvest ever during that time. It was almost impossible to find a dollar and if one was seen to have even a one US dollar note; thieves could 'kill' you in daylight. However, the important point to note here is that the agribusiness foundation had been strongly laid by the old man and all we needed to is to build-on. How do we build on this strong foundation? Infrastructure development is a key component in developing agribusiness; how has the infrastructure development fared in Zambia?

Infrastructure development and agribusiness

Infrastructure development is very important in supporting the agribusiness sector in any country and economy. In Zambia, some basic infrastructure were put in place to promote agribusiness development. Major road pavings were done into the major agricultural areas and provinces. From south to north, ran the great north road which stretches from Livingstone in the south through Mazabuka, Lusaka, Kabwe, Mkushi, and Mpika up to Nakonde in the Northern Province. There is also one stretch that continues from Kapiri Mposhi through Ndola, Kitwe up to Chingola on the Copperbelt. There was another road paved to the east from Lusaka to link with Chipata up to Lundazi while another one goes west from Lusaka to Mongu and Senanga going round to Sesheke. The roads are not the perfect mode of transport to move very heavy bulk goods, therefore, the government developed a rail line from Livingstone to Chingola and another one from Kapiri to the Dar es Salaam in Tanzania. There are several smaller road networks between towns in the provinces. Zambia being a land locked country has no water port except for the in-land one at Mpulungu which links the country to the East African countries of Burundi, Kenya, Rwanda and Congo.

A dilapidated road in one of a farming block in Katete district

However, some of this infrastructure was not fully developed and what we needed was to build on what the first government had started. For instance, Lukulu district of Western province produces a lot of rice in Mbanga area adjacent to the Zambezi plains. Mbanga is about sixty kilometres (60km) south of Lukulu district and about thirty kilometres north of Limulunga. The road from Limulunga though near is impassable as it is flooded most of the time. One can only access Limulunga from Mbanga a few days in November just before the onset of the rain season when it is dry. However, the other route from Mbanga to Lukulu is only accessible by four wheel drives or better off by a tractor. The road from Lukulu to Katunda near Kaoma is gravel and in the rain season, no heavy vehicles can easily pass without challenges. This is very true of other roads such as the Chienge to Lambwe chomba which is another rice producing area in the Luapula province. In the east, the road from Chipata to the newly created district of Vubwi is something one can't dream about in the rain season so is the Chipata – Lundazi road. In Southern province, there are major roads of economic importance that have remained undeveloped for a long time. I have in mind the infamous bottom road which has been on the lips of politicians for ages. Though at the time of publishing this book, some of these roads have been worked on but what needs to be desired is the quality as they are already breaking apart.

The country wasted a bigger part of the ten years of development prospects from 1992 to 2001 because the government then did away with national planning. This government was characterized by corruption and no major infrastructure development was done during this time. Of the major infrastructure development that government will be remembered for was the passenger transport which they turned around and allowed private operators to bring in buses and mini-buses. Before then, it would take one week to board a United Bus of Zambia (UBZ) at intercity bus terminus in Lusaka to the Copperbelt. The buses were inefficient and more often broke down on the way. It would take two to three days to reach Kitwe from Lusaka, unlike today where it can take someone three to four hours. Chiluba was a charismatic leader and he embarked on privatization which wasn't a bad idea looking at how inefficient the public sector had become. However, the act of privatization itself was done in a very corrupt and unpatriotic manner. This will be discussed in a bit more detail on the political decisions made in chapter five. For now we can happily say the first president had laid a very strong foundation for infrastructure development to support agribusiness development in Zambia except to state that he had also stayed too long in power and he seemed to have run out of ideas before being kicked out of power in 1991.

Nonetheless, in the last three years of the patriotic front government (2011 to 2013), we have seen some strides in developing some infrastructure though many of them were planned for by the previous government. The country has observed huge road projects such as the historic Mongu to Kalabo road, the Matumbo road, and many others. However, it should be mentioned that some of the roads needed not to be constructed just now because they are not or will not in the short term add any economic value to the country. The government has indeed gone ahead to construct health centres and hospitals countrywide. These infrastructure developments are important to agribusiness development of any country and in Zambia we needed them so much than ever. However, what needs to be desired is the quality of the infrastructure and the service they are providing to the people.

The basic infrastructure that can support agriculture as well as other businesses are road, railway, water, air and telecommunication. People may wish to agree with me that the country has done remarkably well in the road as well as the telecommunication sectors. The country has roads leading to major towns and districts built by all subsequent governments. The country has only lagged in feeder roads which leads to production areas. The country has a relatively efficient telecommunication sector with three companies offering the services. However, the companies need to do a bit more in improving on accessibility as well as the offering of data services especially in rural areas. The services are also not affordable as compared to countries such as South Africa, Botswana and Kenya. We have fared so badly in the air transport in that we do not have a national airline. People have argued that the country may not need a national airline when it is efficiently serviced by other airlines like the South African, Kenyan, Emirates and Ethiopian airlines. A national airline is very important in developing agritourism as well as the tourism sector. However, the country has two local airlines that services most of the areas in the country but there is still more need in this sector. The water and rail transport are something not to talk about. The country has failed to recapitalize the two rail lines that it has and this presents an opportunity for investments by the private sector. The Zambia Railways used to offer cheap and efficient mode of transporting heavy loads but this is no longer the case. Though we have so many water bodies in the country, we have not developed the water transport. Some questions one may ask is why? It cannot be overemphasised on the the role a sound infrastructure plays in agriculture development and there is huge potential in this sector.

The economy and agriculture

In the last ten years or so (2004 - 2009), Zambia's economy has annually been growing at an average rate of five percent and in instances higher. This has been consistent since 2004 and as expected, this is supposed to translate into reduced poverty for the Zambian people. On the macroeconomic level, there are main variables that will indicate that the economy is indeed recording growth. We have seen this with economic giants like China which has also recorded growth for some time. The only difference is that with our colleagues, if they record growth of even three percent, this will be translated into improved living standards of its people at micro level; this has not been the case with Zambia. The economic growth at macroeconomic level has not translated into microeconomic development at household level. The question many people have been asking is, why? It's unfortunate that I'm not an economist to propound this but I have some vague reasons which I may want share.

To help us answer the above question, let us try to compare two countries: Zambia and Zimbabwe. Zimbabwe has been under sanctions for a very long time from the western countries (Europe and America). When Zambia was going under very difficult periods during the late 80s to the 90s, its people used to get most of the essential commodities like butter, milk, eggs and sugar from Zimbabwe. To taste margarine, one had to have relatives from Zimbabwe. However, I expected Zimbabwe to be getting these essential commodities from Zambia during this period that country has been under sanctions but that has not been the case. Zambia is still getting the margarine, chocolates and many other things including ZESCO transformers from Zimbabwe. If anything, the only thing the sanctions have managed to cripple the economy of Zimbabwe with is the non-availability of adequate FOREX in that country. This has rendered their currency (ZIM Dollar) to be of no value or useless but they still produce those commodities. They have managed to be resilient against all odds. Why has Zimbabwe managed to be this resilient and how have they managed it? The good guess is that Zimbabweans owns their economy and that could be one reason they have survived for over twenty years under gruesome economic sanctions. In that country, most of the factories, retail shops, mines and manufacturing industries are owned by Zimbabweans. What this means is that the money that is generated is retained within the country's economy. Take for instance, the transformer manufacturing company supply our ZESCO (Zambia's electricity company) with transformers and ESCO pays in dollars. The only money Zimbabwe sends out of their economy is what they use to buy raw materials, supposedly to China. However, in the case of Zambia, our economy has a very weak manufacturing base. Anything one can think of in Zambia is imported, including simple things like tooth picks, razor blades, to things like wheelbarrows, bicycles, and foodstuffs. Our economy is only based on copper for bringing in of FOREX. We saw what had happened in 2015 when the prices and production of copper had gone down, literally everything had collapsed.

We all now agree that the gross domestic product of each country is equal to consumption plus investments and government spending and net exports (exports minus imports). This equation can be presented as GDP=C+I+G+Net Exports. In simple terms, it means the country's total wealth is equivalent to what the population can consume in form of social services, balance of payment, salaries for civil servants including buying of medicines for hospitals as well as cash transfers. The I in the equation represents all investments which are the tools and businesses which make products, buy and sells services. This is one important component of the GDP because if a country is not investing, it will end up being like Zambia Railways where even the goods train cannot move because the infrastructure is so dilapidated. Even in a home or company economy, investment is very important because this is what has the capacity to generate future wealth for that economy. Government expenditure or G in the equation mostly does not increase productivity; this basically transfers taxes made from one productive sector of the economy to the other (usually a less productive one). It enables the 'sharing' of resources; in any economy, the government is the largest spender and this explains why some people that could have been doing fine in their private businesses all run to go into government (politics) so that they can have the opportunity to win favours through tenders. For instance, one road project like the Mongu-Kalabo road used more money than the total economy of a town like Kabwe. On the other hand, net exports are what brings in the income or wealth for a country. In Zambia's case, lets dwell more on the net imports. This is because C+I+G normally creates an enabling environment for the goods and services to be produced for sell. Depending on how much of the resources are channeled towards consumption, you may end up with an economy that is based on 'hand to mouth' and less of the resources are being invested. Investment is important because it has a multiplier effect of generating more wealth for the country. Our economy has for a long time tilted towards consumption and in most cases, we have not been able to much it with net exports which has led the country to be borrowing resources every year to support consumption. The country has been a mono economy since 1964 where its many export product has been copper. The amount of goods produced in the country for export has been less than what the country has been able to import. This makes a sad reading for a country which is endowed with so much raw materials such as minerals, forests, land, water, good climate, human capital, animals and so on to be a net importer.

This country is so fortunate that it has so many raw materials, starting with agro based material to mining, tourism, energy, human capital and many other things. Lets pick on a product like copper and look at what other products can be derived from it; they are uncountable! A good example is ZAMEFA, a that makes electric cables in Zambia. They are of superior quality which are of export quality and are exported to other countries in the region. Many Electricians prefer to use our ZAMEFA cables to those coming from other like China because of the good quality. This is what the country needed to have done with literally each raw material that is produced in this country. We might not be able to have our own Silicon Valley because of the low technology that we have but there are certain products that we can produce with the low technology base but quality. The maize value chain is another good example. We have relatively exhausted the products that can be derived from this commodity. Apart from deriving mealie-meal from which we make various foodstuffs like nshima, cornflakes and maize juice (munkoyo and Chibuku), the industries are able to make alcoholic beverages too.

The chain stores that are found in this country such as Shoprite, Pick n Pay, Spar and fast foods like Hungry Lion and many are being provided with a market by Zambians and the money they make is externalised to countries of their origins (South Africa) in form of hard currency. The bulk of the commodities and products that are sold in these shops are imported. The country is just providing the market opportunities and also creating employment for the South Africans. Zambians do not own the economy and it is one reason our agribusiness has not developed to the level of South Africa, Egypt or Kenya. To reverse this scenario, we need to change the way we do business in this sector and it should start with each individual. The starting point is to strengthen trade and manufacturing policies that will favour local industries. Cluster industries owned by locals should not just be in our vocabulary but we need to realize it and empower them. Therefore, this demands that we put in good leadership or strengthen the current ones to help manage the economy well. The country needs to create whatever favourable enabling environment. If the policies continue to favour and supports foreign enterprises, Zambia's economy will remain a big retail trading centre like it is in its current form. It is high time the manufacturing sector is revamped to stir industrial development in all sectors. In the last ten years (2008 – 2018), we have seen the economy shrink at the expense of recording positive 'growth'. Macroeconomic indicators alone are not enough to determine whether a country is doing fine economically, and one reason which has led us to be where we are is lack of consistent good policies as well as high corruption due to ineffective leadership. You are now agreeing with me that politics play a major role in agricultural and economic development. Now that we are in agreement, it is prudent that we together as a team discuss how politics have contributed to agricultural development (positive or negative) in Zambia in the next chapter. Some of the points though have been discussed already in the previous chapters.

CHAPTER 4

Politics and agriculture

In Zambia, agriculture can hardly be separated from politics just like many other countries of the world. Politics drive the agricultural industry and one might argue that this is the situation in many countries in the world however, the correlation is even more pronounced in this country. Our agriculture is highly affected by politics because the politicians literally determine what a farmer needs to produce and when they can produce. Agriculture as we know it is supposed to be a business and indeed in developed economies, it is a big business. In many economies, it provides the raw materials in industries that are used for manufacturing of secondary and tertiary goods. The farmers in this country are trying to take it as a business, but many times the conditions don't allow. There are certain families and industries that have lived and survived on agriculture for ages. Many might ask as how is agriculture affected by politics in Zambia?

Ideally governments worldwide create enabling environments for their citizens and the private sector to do business, that include companies that are in agriculture. The role of governments is to ensure that there is infrastructure and conducive policy framework to support agricultural development. This infrastructure includes things like good policies besides physical infrastructure like roads, rail lines, hospitals, markets, communication systems, banks, markets and so on. Governments formulate policies which enables businesspersons such as farmers to thrive in by doing business in an enabling environment. For example, just after 1991 when the new MMD government was ushered into office, they shifted so much from pro-poor agricultural policies to the elitist. Overnight, agriculture was left at the mercy of the underdeveloped private sector. In my view this was done so sudden and harshly. The country witnessed privatization of institutions that used to support agriculture such that overnight, they were expected to run independently after having been heavily subsidized for close to three decades. Some of the agricultural policies were implemented so haphazardly without considering the level at which agricultural supported industries that where in the value chains. The industry which was highly subsidized from production, marketing to consumption and many of the subsidies were removed so suddenly, all in the new discovered capitalism jive. This is not to say this was bad but the manner in which it was done was not humanely. Some of the institutions did not deserve to be privatized but mere restructuring them in order to revived them.

Unfortunately, this led to the decline in the industrial development tremendously. Initially, the first government had put institutions at each level of the value chain which were supporting the agricultural industry. The foresight by Kaunda was to shift from the mining industry to agriculture as well as other sectors. There were industries such as AFE which were manufacturing and providing agricultural implements to the farming community. This was supplying implements such as ploughs, irrigation pumps, grinding mills and many other implements. Lima and Cooperative banks were also established to provide financial services to the farmers. Most of the banks that currently operate in Zambia are retail in nature and rarely provide financing to farmers especially the smallholders that are perceived to be so risky and are in the majority. Lima Bank was financing all types of farmers while the cooperative bank was mostly targeting the cooperatives movement and its members. For those that were involved in importation and exportation of goods, there was a bank specifically looking at their interests called the Export and Import Bank (EXIM). When it comes to marketing, the government developed the National Agricultural Marketing Board (NAMBoard) which was providing market opportunities for the farmers. This board had established several marketing outlets in all the provinces and they had built holding and storage facilities of which some are still standing to this day as white elephants like the one I saw Imansa School some fifty kilometers west of Luanshimba School after Manyumbi Toll Gate in Kabwe. There are many such infrastructure in the country.

Though maize was the major commodity traded in, there were other commodities such as sunflower and groundnuts. To support the smooth marketing of the outputs from the farmers, Zambia National Service (ZNS) would make sure they grade and renovate all major roads leading to marketing facilities before the onset of the marketing season. This is one facility which was killed by the MMD government. Given an opportunity, this institution should have been smoothly privatized by allowing the private sector get a major stake in the company and allow the government to slowly sell off its stake. They should have allowed the board to invest in vertical as well as horizontal integration unlike the way it was unbundled. With time, this could have been the major supplier of products to ZCBC, the chain stores that could have remained vibrant to this date because that model of operations was not any different from what Shoprite is doing.

The government established several industries in several value chains and those industries were the final off takers of the output from farms and mines. Each commodity supported had at least one factory that was involved in value addition or processing. For instance, sunflower was processed into cooking oil and other secondary products such as soaps and many others products by a company called ROP. This was a fully-fledged industry which was providing thousands of employment opportunities to young Zambians. As opposed to the current situation where cotton lint is exported in its raw form to China and India, the country used to make its own clothes because of the presence secondary manufacturing industries such Kafue and Mulungushi Textiles. These were industries that were supplying various cloth making factories such as the Uniform Shop in Lusaka and other tertiary industries. LintCo was formed to support the production of cotton and it had its own extension system. This has changed hands to Lonrho Cotton, Dunavant to the current NWK. In all the value chains there was both forward and backward linkages. The livestock industry was not left out as the country had Balmoral which still exists to date that was manufacturing vaccines for the livestock industry. Additionally, Zambia Dairy Produce Board (DPB) was buying milk and processing it into various products including butter, ice creams and various other products like milk biscuits which led many of us to be 'disciplined' in homes by our parents as it was irresistible. DPB was the company responsible for offering the market as well trainings to the dairy industry. This was later privatized and sold to Parmalaat, which currently is operating efficiently under the same structures left by DPB though in foreigner hands.

However, with change of government in 1991, we saw a shift in policies. Firstly, I should make mention here that some of the changes that were made were not necessary in my view. Most of the tertiary and some primary and secondary industries were privatized. Many of the industries just needed recapitalization and change of management because nepotism crept into most of these quasi-government institutions especially when UNIP become a 'monster' like we are witnessing with the current leadership. Most of these industries were very important for the continual development of the agriculture industry in Zambia. For instance, companies like ROP was the only industry providing market opportunities to sunflower farmers and once this was privatized, there was no market for the crop. This saw the decline in the production of sunflower, save for some few farmers that remained growing it for their own household processing. This led to Zambia starting to import cooking oil from Zimbabwe, South Africa and later the Far East through East African countries, and Kenya to be specific through the COMESA free trade area. The sad story is that out of all the commodities grown in Zambia, it is only maize which is horizontally integrated to produce finished products such as mealie-meal; everything else, just like copper must be exported in the raw form. Many of these industries should not have been privatized the way they were done because all they needed was to re-organize management and recapitalize then as earlier mentioned. Sad enough, some of these companies that were sold had their equipment removed and relocated to countries within the region such as Zimbabwe. It is worth to note that some of these companies were in the private hands at independence and then the government re-nationalized them. Those that bought these companies could have feared that later with change of government, these might be renationalized again. Therefore, we have seen that politics plays a very significant role in agricultural development in any country. In case of Zambia, the bad policies or haphazard implementation of what was meant to revamp the manufacturing agro-industry and the economy at large led to a 'death' of a sector which was so promising. The greatest mistake the government then made was to demonize anything associated with Kaunda such that even good policies and industries that were functional had to be done away with and industries pulled down to start all over again. This does not mean to imply that everything that was done by the MMD government was bad but concerning agriculture, they did more harm than what the country expected.

Political stability

Political stability is a very important ingredient in agribusiness development in any country. I should state here that Zambia has been very stable politically since it got her independence from Britain in 1964. The country normally does experience some minor skirmishes during elections especially after the reintroduction of multi-partism in 1991. Initially, we never used to experience violence especially in the first three elections (1991, 1996 and 2001) but later on, when certain political parties and political players broke away from MMD, we have seen an increased intolerance towards those in opposition. Serious political intolerance started with MMD when they had the Chawama by-elections before 2001 general elections. The trend has continued where we have seen the selective administration of the public order act (POA). The level of violence that we have observed from 2010 to date can only be equated to what was happening during the fight for independence. The unfortunate thing is that some people have lost property and lives in isolated cases. However, this only happens during elections and once the leaders are sworn in, the country gets back to normal.

To try and unite the nation, Kaunda's government came up with a one party participatory democracy in which all people belonged to one party in the country. This was to try and curb political violence that normally happened during multi-party politics. However, the relative political stability that the country has enjoyed since independence has not culminated into real development that was anticipated knowing that the country has all the resources needed for development. In the first republic, there were some political militants who were called the vigilantes. These were normally youths that were a wing of ruling party UNIPs intelligence and their role was to instill fear in people that they perceived opposed the rule of the government. Unfortunately, in spite of the peace the country has not done well when compared to some of the neighboring countries which have experienced the worst wars such as the genocide of Rwanda in 1994. Currently, Rwanda is developing at a faster rate in many spheres than Zambia including areas of energy yet Zambia has more hydro and solar energy potential. Many reasons can be advanced, some of which might not reverberate well with those that have held political power. Political stability alone is not all is needed for a country to develop. Many think there has been lack of serious leadership especially after the first government. Good leadership as well as institutions of governance that must be visionary and passionate for development have somehow been lacking. Some of the leaders that have been put in leadership roles have unfortunately been there for some selfish motives such as enriching themselves and reap resources from the country using corruption, nepotism and many other negative vices. There has been a lot of extravagance with many of them and have ended up spending on things that were not so important at the expense of real development. Public funds, unfortunately, in many instances have been used for such personal things.

There were some leaders however, that have shown willingness to develop and have tried, but normally what happens is that the country tends to have bad leadership every after we have a good one. Examples can be given; the late Mwanawasa would have made a perfect leader;

this does not mean he was not a good leader but the challenge he faced was that he was elected through a system that was perceived to have been corrupt. It was difficult for him to challenge certain things for fear that he could expose the same bad systems which helped to usher him into office. He tried his level best but there was a limit to which he could fight corruption and worse off, he was fighting a lone battle. Many of his ministers did not buy into what he was doing, as could be seen by their behavior once the man was no more; most of them espoused corruption to the core. After Mwanawasa, the other leader everyone thought was going to change the status core was the late Michael Chilufya Sata; the man of action, so he was called. He was bold enough to tell off his ministers that they were not helpful because he saw that they did not share in his vision and passion to fighting corruption. The country was unfortunate that barely three years in power; the man's health failed him until his unfortunate demise in 2014. He was a practical leader with determination to quickly develop the country. For instance, many still believe that the man was misled on the enactment of a new constitution because this is the promise that he made to the people of Zambia when he was campaigning. Though I was not in government, but it was easy to see that those surrounding him had selfish motives. Just like Mwanawasa, he had very few people that shared in his vision. In a nutshell, this country has experienced relative political stability but the unfortunate thing is that it has not culminated in economic development of the country. We are still amongst the poorest countries on the continent with good statistics such as growth rates which do not translate into poverty being eradication from the face of this country. The common man that had so much faith in the leaders had continued to suffer on the streets.

Government policies

Dr. Kenneth Kaunda was the president of Zambia for twenty seven years from 1964 to 1991. After leading the country for nearly three decades, he needed to have fresh ideas injected into his administration so that the country could move in tandem with the new economic environment that was unfolding. In Europe, one of the largest country in that block, the USSR was divided into several states. USSR was a socialist country and the UNIP government of Kaunda also was leading Zambia with refined socialist ideologies called humanism. This basically emphasized the man was at the centre of everything. However, sometimes political power can be irresistible to the detriment of economic development. Dr Kaunda did not have a vision of Zambia after him and this led him not to groom a leader in time to take over from him. However, when the political change was sweeping across Africa, it led to his letting off power two years before the end of his term due to hardships such as shortages of basic things like mealie-meal, sugar and many other essential commodities. He was defeated in an election that was won by a trade unionist, the late second republican president Dr. Frederick Chiluba in 1991 after a land slide victory. The wind of change that had swept through eastern Europe which saw the fall of the Yugoslavian government as well as the almagamation of East and West Germany into one union country did not spare Zambia. Many belive that this was influenced by USA and its allies like Britain to dorminate with the capitalist ideas. Chiluba was ushered in power and the cabinet he formed was composed of very educated Zambians, most of who had either second degrees with a couple of them having doctorates. Very few ministers had first degree only. Although this is the government which liberated the country from the one party democracy and introduced multi partyism, the country witnessed the worst form of corruption and new culture of governance. Some of the leaders that did not embrace corruption ended up resigning through frastrations or were flushed out of the system. Nonetheless, they had some of the best policies on paper such that certain neighboring countries would come to learn from us but what we lacked was the will to implement. This can be attributed to political will and leadership that embraced corruption and materialism. One fails to understand why they behaved the way they did, but probably it could be attributed to a long time the country was under autocratic leadership which did not allow people to express themselves freely. This was an opportunity for Zambia to correct some wrongs that the first republic could have made. The leaders then could be viewed as those who were in a hurry to develop themselves more than the country. Kaunda did not allow leaders in his government to be involved in business and he might have known that this could easily compromise their integrity. This is exactly what we saw and have continue seeing in subsequent governments where ministers influence government contracts to be awarded to companies that they have shares in or belongs to friends and relatives.

The agricultural industry under Kaunda was managed with so many secondary and in some instances, tertiary manufacturing industries but the new government came on with its own policies which saw the demise of those industries under the disguise of privatization. One of the biggest mistakes Zambia made was to vest so much power in the presidency such that the position is perceived to be like a small god (in my view). The cabinet ministers who are supposed to be his advisor's mostly hero worships the presidency. They fear to criticize him because they can easily be replaced at will and overnight someone will lose the source of good income. The initial economic development which Kaunda started through empowering of Zambians by ensuring that major industries were in the hands of the government on behalf of its people should have been improved on. They had started the industrial revolution in sectors like agriculture and the manufacturing which were flourishing though limping later due to poor management. Most of the companies were being run down because the management then did less to recapitalize them and they had over employed in some instances. The Chiluba government as stated earlier came with different policies of putting the wealth of the country directly in the hands of the people; this could have worked well had they done it rightly but the process was marred with mismanagement and unwarranted 'carelessness'. The MMD's twenty-year rule has been a government of mixed policies. Each leader that took over the mantle came in with different policies and consistency has been a challenge. However, there are certain good policy pronouncements that were made and implemented which led the country to move ahead. For instance, it was the MMD government which tremendously improved the transport sector. They paved the roads which were dilapidated for too long. Overnight, people could travel from Lusaka to Copperbelt the same day, a distance that would take two to three days. People could enjoy exotic fruits such as apples that they could only be eaten if they travelled to South Africa and other foreign countries. It was in the MMD government that we saw investments in infrastructure for market development of our agricultural commodities such as the building of shopping malls, paving of roads and liberating the monetary policy.

Later on, after MMD lost power in 2011, many people had so much hope in the new Patriotic Front (PF) government because their campaign messages resonated so well with what people wanted to hear. However, in my view this has been a government that has been inconsistent with policy pronouncements and implementation. The government started well on infrastructure improvement to see agribusiness thrive in rural areas but this was not supported with a viable economy and disjointed fiscal policies. Most of their policy pronouncements seem not to have been thought through well before being implemented as could be seen by the fact that some policies were implemented and later on withdrawn within a short period because they could not stand a test of time or work well. This indeed, has a negative impact on investor confidence as it affects long term planning. The result has been that the government has only done well in infrastructure development but has fared so poorly economically and on the fiscal policy. Later, this has negatively affected the agribusiness development especially when it comes to value addition which is supposed to be the cornerstone of manufacturing, job creation as well as contributing to economic development. Basically, the manufacturing sector has totally collapsed and Zambia's economy is still anchored on mining with a bit of support from agriculture production. Trading in foreign goods and to some extent service provision. What beats most sensible logic is that they are trading in products such as chips which are imported from South Africa when our own farmers are complaining about the market for agricultural products such as potatoes, for instance. The generation of forex has dwindled so much that the country in most cases has depended on borrowed money to stabilize the local currency (kwacha).

In last quarter of this government's tenure (2015), we however have seen a bit of hope in agribusiness. The government has started formulating policies that favour local industries after receiving so much pressure from its citizenry. The biggest anomaly that most governments made was to think that agriculture was synonymous with maize production. This saw them put a lot of resources in maize production at the expense of other value chains. However, the agriculture minister in 2015 tried to change this with the implementation of the e-voucher system as the way to support the smallholder farmers. This allowed the farmers to have a choice of what inputs they needed to access in order to support their production. There have also been pronouncements or promises by the republican president that the government was not going to be so much involved in the maize business. This is yet to be seen though many people think that there should be a ban on exportation of maize grain but instead allow it only to be exported in processed form as mealie meal. Nonetheless, there has been positive signs in that the government as they only managed to buy adequate stocks from farmers for strategic reserve and allowed the private sector to fully participate in the grain marketing during the 2015/16 marketing season. Indeed, government policies if good promotes agribusiness development in any community and country. The implementation of the e-voucher has led to empowering of the agrodealers who are a very important component in technology transfer to the farmers. These agrodealers have also provided employment opportunities especially in the rural areas as well as taking the productive technologies closer to the farmers in the community, thereby making them affordable and easily accessed.

CHAPTER 5

Agribusiness and commerce

Zambia has received so much support in the agricultural sector in the past twenty-five years than it did in the first twenty-seven-year reign of Dr Kaunda. There have been massive projects run by both multilateral and bilateral partners besides the traditional funders. The support has all been hinged on diversifying from maize production, improving productivity and extension service delivery and promoting value addition. The biggest flop with donor support has been the support channeled towards improving productivity. This has been a thorny in Zambia's agriculture industry for a very long time because productivity has only increased marginally as compared to the potential or anticipated increase.

For instance, with most maize varieties having a yield potential of between 8mt to 20Mt, the average national productivity has remained below 2.5Mt/ha. In 1989, maize productivity was around 1.2Mt per hectare and after the change of government, it dropped to below a ton per hectare due to poor agricultural policies implemented by the new government then which adversely affected agricultural production as well as productivity. It was during the second phase of the MMD government; somewhere around 2002 that the country recorded marginal improvements in productivity which currently stands at the national average of 2.2Mt per hectare for maize though this is still not good enough. The breakeven production stands at five tons per hectare. The question many would want to know is why there has been stagnation in productivity despite receiving so much support in the sector.

Nonetheless, there is no clear-cut answer for this question but so many reasons have been advanced. Some of which borders on the cultural beliefs of its people as well as misplaced policies by the leaders such as;

i. *Lopsided support* – if one must support a pupil to get educated, they should not only support such a pupil with tuition fees alone but they should include other materials that go with getting good education such as books and other learning aids. This is in order to create a conducive learning environment for them. For the support mostly received in agriculture, the focus has been on certain agricultural practices forgetting that this has to be follow a value chain model; all activities that make agriculture tick. Take for instance; the projects that have been giving inputs for production of certain crops have focused on seed only. This includes the support from government which only focused on seed and fertilizer inputs. Crop production needs more than seed and fertliser to be successful. It's only in 2015/16 season that the government's support took a seemingly holistic approach by opening it up to other value chains. Of late Zambians have taken keen interest in farming and many working classes are now turning into weekend farmers especially in the horticultural sector as well as smallholder field crop production. Additionally, there has been an increase, in the recent past of youths taking up farming as opposed to before when farming was thought to be for the tired and retired civil servants and old people in the villages. There has been a 'gold rush' of youthful energetic and educated Zambians taking to farming as full time businesses. This is very good for the industry although not all of them have agronomic background or understanding but they can employ or consult people to help them. This presents an opportunity for the agricultural experts to provide advice in form of extension service delivery and consultancies to those with resources. Authorities are later appreciating that agriculture should be tackled holistically if it has to tick. Some policy pronouncements in the livestock sector such as government negotiating for export markets with other countries have led many smallholder farmers to diversify their livestock rearing enterprises. This is a good example of the government creating an enabling environment for its citizens to conduct business.

ii. *Fertilizer usage* – fertilizers whether organic or inorganic are important to replenish the soils with nutrients. Certain crops such as beans, soybeans, groundnuts and others in that family have the capacity to convert atmospheric nitrogen into plant usable nitrogen or nitrates. Most of the soils in Zambia are slightly acidic in nature. This means certain nutrients even if present in the soils are not readily available for plant uptake. Unfortunately, fertilizer usage in Zambia and Africa in general is one thing that has seen agriculture not to improve for some time because it is under applied in most cases especially in the smallholder segment who are the majority farmers. Zambian soils need on average 300kg of compound fertilizers and the same quantities if not more of top dressing for crops like maize to give a very good yield depending on the soil status. However, the average fertilizer usage in the country is around forty (40) kilograms per hectare. As if this is not too much for the crops, most soils are slightly acidic to acidic; thereby making the applied fertilizer not to be available to the plant. There are very few farmers especially amongst the smallholder that lime their fields to rectify the soil pH. Furthermore, most farmers practice mono cropping and this tends to heavily mine the soils off certain nutrients. There are so many opportunities in fertilizer and lime businesses in Zambia. Though we have about twelve to fifteen companies and traders in fertilizer, most of them have focused on the supplying the government support program called fertilizer input support program (FISP). There are very few companies in this value chain who have focused on increasing fertilizer consumption, worse off, all the fertilizers sold on the market is imported except for about fifty thousand metric tons which is manufactured by a government plant in Zambia for small-scale farmers. The bulk of the fertilizer comes from South Africa and the Middle East. On average the national fertilizer consumption stands at around 400,000 to 500,000 metric tons per year. This can easily be grown to a million and half with the current number of farmers the country has. This demands for prudent use of other agronomic practices such as liming the fields as inorganic fertilizer have the acidifying effect on soils. There are enormous opportunities in the fertilizer business but needs strategic thinking and partnerships. Some companies have started to position production of fertliser in the country in order to remain competitive. However, the market for this commodity has not been fully explored; opportunities are available for those that may think of establishing fertilizer manufacturing plants in the country. On many occasions, fertilizer companies have run out of the commodity mid way the season because of the lead time it takes to bring in the commodity from South Africa or middle east.

iii. *Value addition* – many people think value addition is synonymous with processing but value addition can mean many activities other than processing. For instance, a commodity like groundnuts if it was to be sold in roasted form as opposed to the raw ones, will demand value addition. The cost of the roasted groundnuts will be slight more than the raw ones of the same quantities. Zambians grow a lot of commodities which are unfortunately sold in their raw form. A lot of cotton is produced in this country and all of it is sold as lint. This lint is further exported to other countries like China and India processing. In China, the lint is processed into cotton material and made into clothes which are sent back to be sold in countries like Zambia at very high prices. Zambia has the capacity to make clothes from cotton lint as was case before Kafue and Mulungushi textiles become defunct and were privatised. There are a lot of opportunities in value addition especially the manufacturing sector but the country has opted to be a net importer with most of the finished products coming from South Africa, China, India and the Far East all in the name of globalization. Of the many raw materials churned out of the agriculture industry, the country has only managed to fully add value to maize and soybean coomodities which have been fully explored. ZamBeef is one of the few companies which has explored the full potential of the raw materials it produces on its farm both vertically and horizontally. The company produces soybeans, maize as well as rears cattle and poultry on its three estates. It also buys some cattle from the farmers in western and souther provinces which stocks about sixty per cent of the cattle population in the country. The company is supplying most of the chain store supermarkets with products such as milk, meat, eggs, bread, and flour including shoes. It is also making cheese and of late it has opened several fast food restaurants where several hot meals are served. It has created a lot of jobs directly as well as indirectly in the agricultural sector. Another company that has done remarkably well in value addition is Trade Kings.

This company makes several products using agricultural raw materials that are of international standards. It has explored the regional market and is exporting some of its products to countries like Malawi, Tanzania, Congo D.R., Namibia and Zimbabwe. Though this is the case, there is still abundance of opportunities in manufacturing such as making paste and juice from tomato, juice from fruits such as mangoes, French fries from potatoes, clothes from cotton and many others. Zambia being an agricultural country with many if not all produce sold as commodities, there are opportunities in value addition and manufacturing in that less than ten percent of what is produced is processed. In the next frontier in agribusiness development, the country needs to explore manufacturing and value addition opportunities.

iv. *Vegetable and fruit production* – the bulk of the vegetables consumed in the country are locally produced by mostly the smallholder farmers. One of the hindrances to the fully development of the vegetable production has been the poor road infrastructure that have hindered production of these commodities in far flung areas with better comparative advantages. However, in the last three years (2013-2016) there has been a lot of development in the road infrastructure and this will have an impact on vegetable production in the near future. This is so because most of the markets for the vegetables lie along the line of rail. The vegetable production has mostly focused on leafy vegetables such as cabbage and rape while tomato and onions are the other ones grown. With the opening up of more mines in Solwezi and an increase in the population as well as the economic activities in the DRC, there is likely to be an exponential increase in demand for vegetables. On the other hand, there has been very little done to explore the fruit production especially citrus fruits. Most of the fruits eaten in the country are seasonal such as mangoes and guavas which tend to be available in the rainy season only. However, with good nurturing of the fruit trees, we could experience a whole year-round supply of such fruits due to the diverse of the weather. For instance, in some parts of the country some mangoes ripen during the dry season starting in July. Most of the fruits sold in chain stores such as oranges, pineapples, bananas, apples and lemons are all imported from South Africa. We have the favourite climate and soils which can support the production of such fruits in Zambia. Again, this is an opportunity which we need to explore and further feed into the manufacturing industries such as that making juices and jams. The other thing that needs to be developed is the building of a fresh produce market with infrastructure such as the cold rooms. This will allow the produce to store for very long and ultimately, it will stir up production in this lucrative business. This will also stimulate processing of some of these products as supplying will be quaranteed. The fresh produce value chain in Zambia has a huge potential and it is one of the most lucrative businesses in the agriculture sector.

v. *Livestock and fisheries production* – many Zambians keep different types of livestock such as cattle, goats, chickens and other small livestock. However, there is so much potential for livestock such as cattle in the country. For instance, the national population of cattle is estimated to be around three million but the country could easily accommodate as much as ten to fifteen million cattle. The biggest challenge has been the bad practices such as in-breeding which has reduced the breeding prolific of our animals especially amongst smallholder farmers. Most of these animals are reared in a traditional manner with very poor veterinary service provision which has resulted in the frequent outbreak of cattle diseases that has seen the cattle population being decimated drastically. This also applies to other livestock like poultry, piggery and others.

This is not different with the fisheries industries. The country has been sourcing its fish from natural water bodies. These have been overfished due to unsustainable fishing methods which have seen some water bodies being over fished like the Lake

Bangweulu. As a country, we had done very little to promote the rearing of fish through fish ponds. The country consumed about 150,000 tons of fish per annum of which it produces about 90,000tons both capture and through aquaculture. The deficit is likely to increase every year as natural bodies of fish gets depleted. The difference is imported. Nonetheless, in the last five years, we have seen some activities in this line increasing and this has resulted in having two commercial fish rearing companies established on Lake Kariba. We hope this can be extended to other water bodies as there are more opportunities in this sector. Zambia's fish market is not being satisfied. As this sector grows, we need to see investments such as companies start formulating and supplying fish meal to feed into this industry. Zambia has a fish deficit and there are still enormous opportunities in fish farming though environmental impact assessments (EIA) should be done as this can have negative impacts on the environment such as an increase in cases of malaria if not well handled. Fish provides about fifty five per cent of animal protein to the population.

vi. *Storage and postharvest handling* – the storage and the postharvest handling opportunities have already been talked about at length in the earlier chapters. Zambia's postharvest losses stand at around fifteen to thirty percent of what is harvested; this is too high for the agricultural industry especially ours which is mostly focused on field crop production. Although the government through other partners have established the warehouse receipt system, there is much that is needed to be done to improve the storage of whatever is produced in this industry which goes hand in hand with value addition. If the country can improve productivity to average even as low as forty percent of the potential for the land cultivated in the country, then there will be need to treble its current storage capacity. As agriculture is becoming more scientific and sophisticated, there is need to increase the storage capacity. This means that there are business opportunities in this area. Just to emphasize this point, an example can be cited where all the grains bought in areas like Eastern, Northern and Luapula provinces should be transported to towns along the line of rail for storage and handling. This points to the fact that there are no adequate storage facilities in those areas. This is true for the cold chain supply in the fruit and vegetable sector.

vii. *Market service provision* – marketing of agricultural produce in Zambia has been one of the biggest challenges which have discouraged the production of certain commodities. Generally, most farmers in Zambia depend on the government agencies to provide market opportunities for their output especially amongst the smallholder farmers. It is unfortunate that the private sector has not fully come on board to explore opportunities in this sector. This partly has been caused by the government with its suffocating policies which has ended up crowding out the private sector from participating in certain value chains. They have also been some disjointed activities from non-governmental organizations in this area which has yielded minimal results than anticipated. There is need to explore opportunities that are in this market segment of the agribusiness sector in this country. The enactment of the warehouse receipt system (WRS) is one avenue that presents an opportunity and need to be further explored and developed in helping to provide marketing services to the farmers especially the smallholder farmers who are easily manipulated. This will also demand good telecommunication infrastructure for efficiency dissemination of information especially in rural areas. Currently, there are literally no companies that are providing a platform for this service and farmers rely on their mobile phones to communicate with friends near market areas. The farmer's union tried to create a platform they were calling farm prices but it was not being updated so often and became very unpopular. We need private investment in this segment.

viii. *Disposal of waste and environmental management* – everywhere one looks in urban areas, the probability that their eyes will see a heap of garbage is more than fifty percent. The biggest nuisance has been plastics, paper and leafy materials from agricultural produce. The government through ZEMA should declare garbage in Zambia a disaster and they should ban production of plastic carrier bags like our colleagues have done in Kenya and other African countries. This is clogging all drainage systems in the towns and reducing productive lands in rural areas. Large cities like Lusaka have run out of land to be used for disposal of the waste. Nonetheless, in other countries some entrepreneurs/companies have taken advantage of this and developed recycling enterprises. This is a great opportunity in this country to even make manure from waste farm produce and recycle paper and plastics. There is great opportunity for those that want to invest in waste disposal and management (at the time of sending this book for publication, the government had banned the use of plastics less than 30microns). In the next two generations, some of our land may be declared unusable because of disposition of plastics and this will start seeing itself in the food value chain.

In a nutshell, there hasn't been a lot of development of industries that are related to agriculture in Zambia especially at tertiary level. There are several reasons that can be attributed to this but the bulk of it hinges on lack of access to finance, lack of political will and poor supportive infrastructure such as power, roads and inadequate technologies amongst many others. This book will try and elaborate on these points in the later chapters especially when we try to discuss some of the practical examples of commerce in chapter six. The sector if well developed and integrated in commerce can be the lifeblood of Zambia's economy. The basic infrastructure that supports commerce such as financial institutions, government policies, education and many other are in place for this sector to thrive and all that is needed is to improve on them and implement them fully. As earlier stated, there are many opportunities in agriculture, manufacturing, power generation, tourism, health, education, construction, mining, financial, communication and service provision. Notwithstanding the development that has taken place, the country can still do much better and this can create not less than five million direct jobs in this country within a short period of time. For those with resources to invest, look no further than Zambia. The people are so friendly, the environment is good and the policies (both fiscal and economic) are relatively stable and favouring foreign direct investment. Since 2004, the economy of the country has annually been growing at between 4% to 6%. There is no country in the whole of Southern African which has a better endowment of factors of development and conducive environment than Zambia.

CHAPTER 6

Agribusiness in practice

In this chapter, I will endeavor to share with the you some of the articles I had been contributing to some daily tabloids in Zambia in a column which was highlighting Agribusiness development for over seven years. The idea is to try and provide the readers with some agronomic as well as agribusiness information about some major crops grown in Zambia, and some decisions that have been made both for and against some industries. This will make you appreciate and understand well the strides made by farmers and government in this region to develop agriculture. The articles will try to highlight the practical application of entrepreneurship in the agricultural industry. In the same vein, I will try to share with you some of the best and worst decisions that the politicians, leaders and businessmen could have made to develop as well as kill the industry in a summary. I should state here that the analysis is based on my personal belief and views, where possible I have tried to quote some great scholars as well as leaders. These articles were written between 2012 and 2018 just before publication of this book. I was compelled to compile them in form of a book due to the popular demand from the readers.

Political will and value addition

Being a landlocked country, Zambia has eight neighboring countries surrounding it. In the region, it commands better comparative advantages in as far as potential in agricultural production as well as agribusiness development is concerned. Its land can support literally anything one can think of growing in the tropics except for production of crops grown in the temperate regions like apples, which can be grown with difficulty. The country has forty percent of water found in the SADC region, good fertile land with abundance of wild fruits. The forests are rich with a lot of wild life, timber and 'sweet' grass which can support rearing of any herbivorous animal. The climate is ideal for agriculture and top of it all, the country has enjoyed peace since it attained its independence in 1964 from the British government. A lot of praises on potential has been showered on Zambia with minimal realization of wealth in the last fifty years of its independence. The people fail to understand why the country is still poor with so much resources to use for development. There is so much potential that this can only be understood by those that have lived in the country for some time.

The biggest problem that has been identified is mainly poor leadership or lack of it in most cases. Literally everyone is potentially a politician of some kind and in many instances, they are or waiting to aspire to lead the country. These are very good at talking and do less or nothing in terms of implementing what is talked about. Politicking is health but if it comes at the expense of economic development, it then becomes toxic. They are all seeking the comfort zone and once it is attained, they even forget the people that helped them to get to the comfort zone and yet they claim to be working for the people. I tend to agree with one former minister, Hon. Masebo who commented on a TV interview and she said, "this country is not poor and please don't allow me to believe that it's our 'brains' that are poor". To some extent we have allowed ourselves to be poor in the midst of abundance because we have somehow tolerated mediocrity. Partly this has been caused by our culture which embraces the 'I don't care' attitude; if it doesn't affect me, then it is not my problem. This unfortunately, is not just with the attitude with common people but our leaders as well. This can be seen by some of the unpopular decisions that they make. It could be a cultural issue though people are now realizing what they are owed and are standing up to challenge those that claim to lead them.

What can one make out of the fact that a lot of people in the world think Victoria Falls is found in South Africa? What did South Africa do that Zambia has failed to do to make people believe they jointly own the Victoria falls with Zimbabwe? We always claim that it is too expensive, for instance, to advertise in renowned international media platform such as CNN but if one has to ask how many latest vehicles the political leaders buy every year for themselves, one will realise where our priorities are put. The engineering departments of the ministry of works and supply have been defunct because there are no vehicles to repair. The government buys new vehicles every so often for the same people, yet many Zambians have been driving *salaula* (second hand) cars from Japan bought as far back as ten years from the time they were manufactured but are still in tip-top condition and running while ministers change vehicles every two years if not every year. On average these ministers have not less than two vehicles and more in some instances. It would not be wrong to conclude that the resources are being abused at the expense of the suffering majority. This scenario is not synonymous to Zambia but many countries on the continent are facing the same challenge with an exception of a few.

Not to veer so much into the 'wilderness' of politics, let us get back to the topic at hand. Let me start by stating that many people are disappointed by our governments' slow or lack of policy implementation to realize value addition in agriculture. The problem with most of our leaders is that when they talk about agriculture, what comes to their minds is maize production. Our government (PF) promised that within a short period of time after being ushered into power, Zambia was going to see drastic changes in agriculture but that hasn't been the case. We are still seeing the old ways of doing things but only changing the names. There haven't been any real progressive policy pronouncements, let alone implementation in agricultural value addition which is the vehicle for any industry's wealthy creation of a country. Real poverty alleviation lies in wealth creation and this can only come about due to increased productivity, value addition, manufacturing and effective trade anchored on good policies and implementation plans with strong monitoring framework. At the time this article was being written, it had been two years since the new government was ushered into power. They claim that time was not enough but we could at least have seen some progression in that direction – good policy formulation to start with. For instance, we had already seen the direction in which the MMD government was driving to in 1992 barely a year of being in leadership.

However, there have been a lot of good pronouncements on infrastructure development like the Link Zambia 8000, which is an excellent project once completed but people expect much more than that from its government. The president means well but many Zambians are afraid that some of his generals (ministers) are not with him in sharing-in his vision. This makes a lot of people to wonder as to whether a bulk of them are in government to beat the riches of some great political leaders that have built their empires for a long time through using underhand methods of enriching themselves. The country has a lot of raw materials in agriculture but for God's sake, it continues to export them in their raw forms without any value addition at all. This has been limiting real development especially where job creation is concerned. Take for instance; the country is a net exporter of vegetable or cooking oil as it is fondly called. One would think the government would have deliberately come up with measures to limit the importation of cooking oil into Zambia because the country is self-sufficiency in manufacturing of edible oil[4]. The country may have problems with availability of oil for a season but after that, it is always self-sufficient in oil production just like the case is with wheat production. The companies importing cooking oil would have been forced to come and open up processing plants here in Lundazi, Mkushi, Monze, Petauke. By so doing, we will get more

---

4 At the time this chapter was written, there was no ban on importation of cheap cooking oil from East Africa but later in 2015, it was implemented and lifted again after prices of oil soared

benefits than importing palm oil as a finished product from Malaysia. In 2011/12 marketing season, the cotton farmers were subjected to a lot of misery in as far as cotton marketing was concened. Many of them made losses because the commodity prices were crushed due to over production. Suppose we had secondary processing factories in the country, the two big companies who are the major players in the value chain would have paid economic prices to the farmers. Some technocrats might claim that it is not government's role to force the private sector to be innovative, but the government is the major stakeholder in industrial revolution through creation of enabling environments for the private sector to prosper. If the MMD government wasn't reckless with its privatization drive of public institutions, products like Tarino, Kwench and other brands of soft drinks would currently have been favorably competing with international brands such as Coca cola and Fanta today. The problem is that we sometimes leave such important decisions to be made by political leaders like the president. The president can't do everything by themselves but they need guidance from the technocrats. However, the president also must be accommodative enough and take to listen to advice from other people, which sometimes is not the case in third world countries.

There are of course certain individuals that I have admired. For instance, it's not a secret that a certain minister was removed from the ministry of mines (2013) because he believed in re-introduction of the famous windfall tax. The minister had stood his grounds and advocated for windfall tax even when everyone else including the president had said that windfall tax was not the best, but this is the same tax regime that the late president Mwanawasa's administration used to get the most benefit from the mines. Mind you, the mines were mostly sold at the least value due to the conditions in which they were at the time, though the buyers were able to make more money within a short period of running them. Such are the type of leaders the country need to have who can stand their ground if they know that what they are fighting for is in the interest of the country. What he only forgot to do was to resign after being frustrated with his resolve to have the windfall tax re-introduced. The country needs to have strong leaders that espouses good policy statements in value addition of agricultural products. It's unbelievable that we export soybean cake and later import feed stock, why? Why should the country be importing birds (chickens) from South Africa, Brazil when these can easily be reared in Zambia? There is a lot of political rhetoric as regards value addition; this country needs implementation of policies which will lead to attainment of tangible results that can be seen. We are not saying this government will achieve all this in five years but there is always the beginning. Almost half a century in mining, the country is still exporting concentrates at the expense of finished products like those manufactured by ZAMEFA!

With all the mangoes that are grown in Eastern, Luapula and Western Provinces, it is disturbing to report that the country does not even have a single factory that is making juices and other products from the mango fruit. Many people have observed the potential in this sub-sector but they have not dared to venture into it because they cannot access affordable finances for instance. There have been some 'factories' in Matero and Zingalume that had been making juices in their backyards using simple methods. Those people are driven to be doing that in unhygienic factories because of the prohibitive conditions to establish factories and other enterprises whilst they had seen a yawning opportunity. If those people are properly harnessed with a bit of training and exposed to cheap capital financing, they have the potential to reach the levels of Silvia Caterings, Coca cola and Barum Beverages. This has been demonstrated by Trade Kings through the drink they make called Maheu, which has given great competition to coca cola. The government needs to have facilitated such innovative people with entrepreneurial minds to access resources so that they can open up small proper factories where they could have been making those juices and train them in food safety. They have demonstrated their innovatives. The challenge of accessing finances to run legal businesses from banks should be facilitated by the government through monetary policies. These enterpreneurs should not be gagged but sit them down and engage in meaningful consultations. The country is tired of the rhetoric about value addition in agriculture and mining by our leaders. Since 1991, they have been singing the same song with no meaningful progress but expecting different results. Political will is indeed important in value addition. We are not asking them to open companies on behalf of the people that will be adding value but we are asking them to formulate policies that are conducive and supportive of value addition. One may ask why we are so concerned about value addition when we can have square meals. We can see the importance of value addition and what it is ccapable of doing to the economy of this country in the next article. Value addition can contribute massively to wealth creation and economic development of this country.

The importance of value addition

One day when I jokingly told some farmers to add value to their recently harvested groundnuts, they did not take this kindly and asked of how they were going to do it without machinery. In their view, value addition is processing and for this, they were looking at owning a hammer mill for them to grind it into powder before selling. As we have earlier discussed, many people understand value addition as being processing only. What many don't seem to understand is that processing is just but one method of value addition. The oxford Dictionary defines value as 'how much something is worth in money or other goods for which it can be exchanged for'. Value addition means an array of activities that can be done to improve the marketable value of a product or service.

Let us give an example that is normally used in our everyday life. Take an example of a beautiful lady who has just woken up from sleep; she has not yet bathed and cleaned herself up. Give that lady some good one hour in the bathroom applying all the makes-up and dressing up nicely. Would you think she will be as valuable (in terms of appearance) as she was before going in the bathroom? You might ask what processing activities she has undergone whilst in the bathroom? Nothing! No plastic surgery but what she did to herself was just to take a warm shower, apply the lotion and the make-up to add value to her looks. This is the same context we talk about when we ask our colleagues in agriculture to add value to their produce. Of course, this does not imply to say processing is not value addition but it is just one method of value addition as stated earlier. The lady can easily go for breast enlarging and facial changing like some celebrities in some countries have done to transform their physical appearances, and that is processing because there is complete transformation of the physical appearance of certain body parts. However, we need to state that processing has more potential to add more value than probably sorting and grading. It also has huge potential to create more job opportunities than simple sorting and grading of commodities.

In this case, what I was trying to put across to the farmers was that for instant, to sort out their groundnuts before selling is value addition and this will make their produce fetch a better price, and ultimately they would make more profits as opposed to how they where doing it. In this case what they normally did was to mix both the normal nuts and the shriveled ones in the same bag. This was done in order to increase the volume or quantity of the product as their harvests are normally too small for them to make any money. Their productivity was normally around 0.4Mt/ha as opposed to the potential of 1.5Mt/ha. This is one grave mistake most farmers have been making and it affects their pricing. In business, it is a known fact that profitability is not a function of only one parameter – price, but it is a function of price, volume or quantity and quality as well as efficiency and to some extent consistency. One would rather sell a twenty kilogram tin of good quality well sorted nuts than selling a fifty kilogram bag of nuts that have nuts mixed with shriveled, rotten and good ones. One risk selling that bag which might contain about thirty kilo of very good quality nuts at a lower price than the twenty kilo tin of sorted nuts. The point I' m trying to drive home is that whenever selling agricultural produce farmers should learn or make it a habit to sell quality through value addition. This can be in form of processing (transforming it physically) or sorting and grading.

When the government officials ask farmers to add value to their produce they don't necessarily mean they should always process their produce. Washing tomatoes before displaying them for sale including packaging them are other forms of value addition that farmers ned to practice. For those that can process, they are urged to do that because it tends to give them more value than primary value addition. Katete District Women Development Association (KDWDA) are involved in processing of cooking oil and one day I stopped over to buy cooking oil which they were processing from groundnuts and sunflower, the quality was not any different from that which is sold in some of the chain stores in Lusaka. If farmers' associations and cooperatives countrywide could emulate and adopt the business model of this organisation, the country will have a lot of SMEs and value will be added to farmers produce at different levels. The association has created some jobs and are improving the nutrition status of the people as well as the economy in community they are operating in. It was unfortunate that during the time of my visitation, they were not processing because there was no power. The power generation company in Zambia has let down many would be investors; they should think of utilizing the abundant sunlight. Industrialization that has started in rural areas should be sustained by the efficient supply of power. Indeed, the producers need to add value to their produce. However, we need to own and protect the industries that are adding value if the resources are to be kept within the country. For instance, on the evening of 31st January 2016 Zambia national team was taking on the mighty Guinea in the CHAN second quarter final match after the Mali versus Tunisia game. The game was to follow the triumph of Zambia's best female boxer Catherine Phiri who defeated the Mexican champion to grab the WBC world championship belt in her own backyard in Mexico. Mexicans are known to be very resilient boxers. This was a great victory for the Zambian and Africa at large, and many hoped the young girl's effort could be recognized by government. Her hard work and discipline has really helped market this country just like the late Lottie Mwale did and we look forward to a day when Lennon, the ring announcer will be doing his thing in style here in Zambia. I will not be shocked to see a horde of banks hosting her knowing that she has a couple of dollars to bank (this actually happened when she returned).

As I was streaming on the internet one day, I bumped into an article that sent me crawling to my rocking chair to put my ideas to paper. The article was about the sale of one of the great milling company on the Copperbelt province to a South African company for a paltry $27.5million. I should state here that I did not know the exactly value of the plant then, but the two times I had been to that factory, it's a modern piece of investment in agribusiness. The milling plant which used to be part of agribusiness company that was owned by Commonwealth Development Corporation (CDC) jointly with the government until the late 90s. It was later sold off to Zambeef group and the plant to the current owners (as a family business). It was a marvel of an investment in value addition which I feel every corporate farm should think of integrating their activities if jobs and wealth are to be created for the country. MDC as it used to be called was a vertically integrated agribusiness enterprise. The coffee grown at the farm used to be processed at the same plant and sold as a finished table product ready to be consumed. So was the maize, wheat and other products. My last visit to the plant was in 2009 when we were looking for a market for the produce of smallholder farmers of Luapula province where I was working then under one of the projects jointly funded by the Finnish and Zambian governments. It had diversified its product line as they were able to produce and sell things like macaroni and other vertically integrated finished products. At one point, I used to think that MDC was going to follow in the footsteps of Zambeef until when CDC pulled out and I was quite taken aback because I had so much confidence in them breaking into foreign markets because of their production efficiency.

The reason the commerce ministry should have regulated that sale of that plant was that there have been a lot of industries in Zambia that have been taken up by South African companies. The country is like an extension of that country's commerce industry as we no longerown the trading business because all the products that are sold (85%) in most of the shops are coming from South Africa, with about five (5) percent from Egypt and an equal proportion from other COMESA and SADC countries. Suppose overnight we have trade embargo from South Africa (I do not anticipate for that), what can happen to our economy? South Africa literally is controlling our economy as far as agro-products and trade are concerned. One wonders why the country has allowed the situation to be like that. They have opted to maintain the plants in their country as a way of creating jobs for their people and only export finished products to Zambia. No one is against FDI but we have Zambians that are effectively running such plants for instance, and these need to be supported if the manufacturing sector is to boom.

The government could have easily enticed Zambians to take up this plant as major shareholders in partnerships with the South Africans that had exhibited interest in purchasing it. This could easily have been regulated about the maximum shareholding they could have bought in that business and floated the other shares on the stock market in order to empower the local people. These are examples of deliberate policies that should empower Zambians to own the manufacturing base. It is unbelievable that already established industries are easily being given away to foreigners because of policies that are too liberal. If the South Africans want to create a conglomerate in Southern Africa, they can be encouraged to start up green investments unlike being 'vultures' of what is already established. Zimbabweans have managed this far to withstand the sanctions by the western countries for a very long time because they own that economy. They might not have their own currency but the bulk of the manufacturing industries are in the hands of Zimbabweans or are based in that country though owned by foreigners. I strongly objected to the sale of that great agribusiness firm and hoped that the powers that be could have seen the sense in it. Zambians have been sleeping as a country for too long a time; it's time to wake up. The country needs to protect what it has established. There is no point in preaching to establish cluster industries if it cannot maintain them as its own. Countries that have developed have invested massively in value addition and manufacturing. Developing the manufacturing sector through growing the SMEs is the only way jobs can be created for the country as well as wealth. Therefore, the importance of value addition in economic development is non debatable in an economy like Zambia. It is a must that Zambia should strongly embrace value addition and make it a culture for each sector especially agriculture. For people that want to invest and get quick buy-ins, investing in value addition for many of the enterprises in Zambia is viable. The manufacturing sector, especially the agro based has so much potential to create jobs as well as wealth for economic development of the country.

Agro- enterprises and competitiveness

Many people have endeavoured to start value addition enterprises at micro level by establishing SMEs in the agro-based industries but have tended to fail. Many may have wondered as to why Zambia's agro-based industries are not as developed as it is supposed to be. The country has almost all the major resources needed for its economic development as stated in earlier chapters in this book. Most of the solutions to developing our agricultural industry in this country are known but we have sometimes made decisions that are not favourable for that. Well, let's remind ourselves again about some of the decisions we need to make to transform our agricultural based small and medium enterprises (SMEs) into fully fledged industries or companies that can compete favorably on the international market.

Several times, our leaders have reminded us time and again that at independence Zambia's *per capital income* was higher than that of Malaysia and South Korea. Today these two countries are more developed than us in all spheres of economic development except to say we still have more natural resources than them. In 2010, Botswana's GDP per capital was US$14,000 while in the same year; it was only a party US$1,500 for Zambia. This country enjoys a comparative advantage in many areas such as natural resources, human capital, population and so on but it is still less developed than Botswana in terms of the quality of life of its people. Some may argue that the population of that country is just under three million people compared to Zambia's seventeen million. The question is why then, when that country has slightly under less people with less than a quarter of the natural resources than Zambia has? Zambia make better policies than some of her neighbouring countries yet it is overtaken on terms of development. Unfortunately, this ugly scenario is even more pronounced in the agriculture sector. Let me not be a lecturer of business strategy or economic development, but allow me to share a few success stories of some projects in agriculture or rural development within the country.

There are so many agricultural related enterprises that we have tried to develop but unfortunately have failed. I choose not go back to delve on the foundation that was laid by our first political leaders after independence but lets us try to discuss the efforts of post 1991. The reason is simple. This is the period when the country made some of the decisions that were not popular. Many remembers some of the key industries that were absorbing the agricultural products and producing some raw materials (Mwinilunga Pineapple Factory, Kateshi Coffee, Munushi Banana Scheme, Mulungushi and Kafue Textiles, Mpongwe Milling, just to name but a few). It is true that the policy makers of the first republic even though they were less educated, were more effective than those of the post 1991 with all the educational credentials. The country has so many mangoes in Luapula, Eastern, Western and other provinces that just go to waste during the rainy season. None of these fruits are found being sold in the supermarkets but the people will instead buy exotic mangoes from South Africa and bananas from Costa Rica. Roughly, twenty to thirty percent of the mangoes just rot while the people drink mango juicy from South Africa, Egypt and many other countries. Like I have mentioned almost in all my articles throughout this book, I am a strong believer that markets stimulate production though industrial development comes with deliberate policy implementation. In the early 90s, the Danish or is it the Norwegians and WCS provided us with some funds to set up projects as alternative livelihoods in communities that live near or around game parks and game management areas (GMAs). The project was centered on developing alternative livelihoods in agriculture and non-timber forest products (honey). This was to deter the citizens from poaching in the game parks as a way to conserving and managing wild life.

Through judiciously following the value chain analysis, poachers were organized into groups or cooperatives and they were trained in various agricultural productions of various commodities based on the comparative advantages of their areas. The former poachers were trained in skills to cultivate groundnuts, soybeans, rice and on how to keep bees for honey harvesting. The key factor to note was that they established processing centers which were the nodes for market establishment. These centres have evolved to date such that a company has been born from this effot. Many retail shops and supermarkets in Zambia are now stocking products from this company which are branded as 'its wild'. From a mere NGO a company that adds value to various agricultural and forests products has been established. This company is providing several job opportunities to Zambians as well as contributing money in form of taxes to the country. In return, it is providing market opportunities to farmers that would have struggled to find such without this innovation. Products that are manufactured by the company are of international standards such that it is able to export to other countries in the region such as Malawi, South Africa, Congo and many others. Some products like honey are been exported to Europe as well.

At the time of writing this article (2012), this company was establishing a central operations center in Chipata as they initially used to operate from a periphery town of Lundazi. This SME is now a fully operational company with various departments such as the origination who are responsible for offering extension services to the farmers that they buy from, the production department that are responsible for various value adding activities to transform raw agricultural commodities into finished products of international standards such as peanut butter, well packaged and branded rice, beans and honey. They also have the marketing department responsible for distribution and sales. If one is to find out how many jobs this enterprise has created, it should be more than hundred permanent jobs with over two thousand seasonal jobs. This is a company that has been formed from an agribusiness related project with the view to saving wildlife. They have not only created jobs in the agriculture sector but in the tourism industry as well. Without this intervention, some game reserves could have been closed by now due to poaching. The only unfortunate thing I have noticed with us Zambians is the culture of inferiority complex against our own products. Honestly, if we don't buy our own products so that we can provide feedback to our manufacturers so that they can improve on the quality, who will do that? If we do not believe in our own product no one will and those products will fail even before they reach the market.

The other very important project that has left a mark is profit. This was a project funded by the bilateral donors from USA to help establish markets for smallholder farmers through various interventions including financing. One company in the name of MUSIKA has been registered as a training company or service provider born from this project. There are some other companies that have been formed in similar ways such as ZATAC in the service industry to provide consultancy services. It used to offer various capacity building services such as agriculture enterprise development and financing. It has continued with its business line as a registered company. I am laboring to go all this far because I have a strong feeling that those people that fund projects especially agricultural related should have deliberate clauses in their Request for Proposals (RFP) or contracts that if the funding involved is colossal and for a long period; the implementing organisation should at least establish a tangible business enterprise by the end of the project/program. It is not only enough to report that so many people were trained, so much volume of commodities was sold and so many people are now leading a better life than before the project. Most of these companies that you are seeing coming to build our schools and roads from China started as SMEs.

While working for one of the agricultural input supply sector; we used to order the bulk of our products from China. There was this Xinhua (not real name) was responsible for the African market. After making orders for about three times with that company, and when we wanted to place for the fourth time, the gentleman asked us to transmit the money in a different company name and we later learnt that he had established a plant with a few friends that he was working with. It is establishment of such enterprises and business which will make our economy grow through creation of so many jobs that will be contributing to growing our GDP and the economy in general. The development of the economy and creation of industries and jobs depends on harnessing SMEs such as these. It is the governments role again to create such environments through various policy frameworks.

The panacea for real diversification

A lot has been said about the need to diversify to agriculture from mining and within our crop production. At some point, we have sounded like a 'skipping record' playing by repeating the same message but with little progress. Sometimes we sound as though we don't know the factors that can lead to successful crop diversification or is it that we want to hide behind the terminology while keeping doing business as usual?

Zambia is a maize country and it has been growing it for a very long time, though we are still growing it like amateurs. The productivity is still too low though the country produces enough to feed itself and have some to export to the region. This is due to the sheer numbers of farmers that are involved in the value chain and the large area of land under cultivation. The current level of maize production in Zambia (2013) under the smallholder farmers is not profitable unless one either considers to add value or export it to countries like DR Congo where a bag can fetch as high as K250. This does not mean maize farming is not profitable, but what has made it to be unprofitable in Zambia is due to the low productivity. What can Zambia do in order to diversify into other profitable value chains or to make the maize farming profitable? The country needs to venture into value chains which are commercial in nature. For instance, the price of cocoa is selling at US$1,400 per ton on the international market. This does not mean that the country can abandon maize production all together and venture into cocoa production. The country might not have the comparative advantage to grow cocoa due to the climatic conditions that prevail. Nonetheless, there are certain things that the country needs to do for it to diversify. One of the most important things is to have the markets and iron out the structural constraints which hinder attainment of perfect market conditions for agricultural products. We have relatively tackled well one area of improving the marketing infrastructure (roads) in the last four years of the current government. If you are in Lundazi and throw a projectile, it may land in Malawi; when in Mansa and do the same, it may land in Congo DR; the same with happen with one in Mongu with

Angola. All those countries we have mentioned besides many others are huge market opportunities for Zambia. Our brothers from Angola and Congo DR consume a lot of cassava but we have large quantities of cassava grown in Luapula and Northern provinces, yet our farmers' cry that they lack markets for the commodity. In order to move 30Mt of cassava to Congo; one should either go through Serenje, Kapiri and Kasumbalesa before landing it in Congo. You do not have to think of taking the same volume to Angola because there are no 'roads' that will permit that volume into that country. What I imply is that transport infrastructure is an important ingredient for having perfect markets. Indeed, we need to work on the communication infrastructure which must go beyond roads but cost of communicating as well for easy transmission of marketing information. Our marketing information systems need to be improved as well because information is a key component of marketing.

Secondly, this is a country which has been sitting on potential for over fifty years and yet little has been done to explore it to realize the economic benefits. In the past three years (2014-2016), it came to the fore that there is need for serious investments in the energy sector. The country has received so much good will from the mining investors in the last eight or so years. However, the mining sector is not operating at full capacity because the energy sector is failing the industry. It is difficult to imagine what will happen once big projects like the Luena and Milenge sugarcane projects are fully operational in Luapula province. It is well known that energy is the backbone for manufacturing, value addition and mining sectors. There is need for seriously investments in energy generation if agricultural commercialization is to be fully exploited. Thirdly, it is possible to have all the infrastructure and raw materials in place but without the skilled manpower, it is as evil/bad as having those *Sangomas* (traditional healers) who prescribe that having canal knowledge of miners is the panacea for curing AIDS. There is need for investment in education sector as well; and by education we do not mean just building classrooms and houses for teachers but there is need to go look at all activities that comes with provision of good education. There is need to critically look at the curricula in primary, secondary and tertiary schools to ensure they conform with current needs of the industries. I tend to agree with those that have a view that our education system is tailored to training people to be employees. Therefore, as the country tries to sell the investment opportunities in the agricultural sector, there is need to invite investment which will create market opportunities for other value chains other than maize. There is agent need to look at other lucrative value chains such as the cashewnuts and macadamia production for instance. I am well informed that the price of cashew nuts per ton at the world market is higher than that of copper of the same unit, yet it is easy and cheap to produce a ton of cashewnuts than copper. The barriers to entry for the cashewnut value chain is a hundred times less than that for copper. The country has cashew nut trees and plantations which have been abandoned in Western province that was set up by the UNIP government. Mozambique which established it cashew industry just recently, is doing much better is this regard. The country needs to revive that plantation in Western province.

On the other hand, the government can deliberately invite investments in the energy sector by treating all those that want to invest in the sector as kings because it is one ingredient that the country needs more than copper. They can give them incentives by allowing them not only to generate but distribute too, using their own infrastructure they can set up. This can be done by revising the energy policy which only allow investors in this sector to generate and sell their power to ZESCO. In all that the country is doing, it should not forget that it is the best equalizer. The entire panacea for crop diversification lies in having good policies through real political buy in by our leaders. Formulating and implementation of good policies will drive real diversification in agriculture and other supportive industries. The political leaders need to be taken to work and not treat them as though they are doing a country a favour. If diversification in agriculture has to be promoted effectively, it will lead to job creation in the industry and ultimately economic development for the country. It is evident that jobs are really wanted in this country as the unemployment rate is very high as well as underemployment. If more jobs are created, there will be more taxes paid to the treasury and this will lead to economic development of the country as they will have more resources to invest in other social sectors for the well being of its people. The question that may beg answers is how can jobs be created in agriculture besides what we have highligted?

Real job creation in agriculture

Zambia's population is youthful and this is a great opportunity that the country has for economic growth because most of the people are energetic and serve as a good reservoir for the labour market. Quality of this labour pool could be another issue as education standards provided by the public institutions have tremendously gone down in the last few years. Other countries like the United States of America (USA) have a population with more baby-boomers and you will find that people as old as seventy-five years are still in employment. This has its own merits and demerits but I think the demerits outweigh the merits more especially with those jobs that are more strenuous. If countries like Zambia can churn out more skilled labour, it can export to such countries and earn revenues that can contribute to the GNP. In China, the government has done away with one child policy because of the aging population. It has projected that if they continue with this policy, labour availability will be a very big problem soon though there are over 1.3billion people. You may also realise that there are a lot of foreign direct investments in that country because of the availability of cheap labour, and this is a competitive edge they have over countries like Zambia.

However, this is not the case with our country because out of the employable youths, only a few are in formal employment. Most of the youths have taken to trading, which in itself is not a bad idea if only it can be formalized so that they can contribute to the treasury. I have even lost track of the unemployment rate in this country; it could be over eighty percent of the eligible population. That is too high a number for a youthful country with a small population but with abundance of raw materials. This scenario can change within a period of five years if the country can put its acts together and implement fully, whatever good ideas or policies it formulates. The country needs to create not less than five million jobs to cope with the ever-increasing unemployment levels. This is possible because there is vast natural resource base that have not been well tapped into but are being unsustainably exploited by foreigners at the expense of the locals. The country is only benefiting crumbs from such God given treasures. We have allowed ourselves to be poor and as rightly put by the late president Sata who said that God will not take kindly Zambian's inability to efficiently and effectively use what has been given to them.

Many times, you have heard the word potential in reference to the Zambian scenario. This word has been used more than often, especially from politicians and those charged with the responsibility to be making investment decisions on our behalf. Every time a politician or civil servant with political inclinations, for instance, District Commissioner appear on public media such as the nation television, the first word that will be misapplied more often in relation to economic development in this country is potential. Many of us have even lost track of what the word mean because of the context in which it has been misused. Indeed, Zambia has massive agricultural and industrial potential in the region that can only be compared to none. The country has the water, the land, the sun, minerals, forests, good soils, the people and on top of it all; the market opportunities in the region. Just imagine how much cash the country could have made had it harnessed fully the agricultural potential in 2015 when it was the only country in the region with surplus maize. Putting all the maize producing countries in the region, Zambia was the only country that had a surplus. It is not a secret that Zimbabwe had her eyes fixed on Zambia, so was Malawi, Mozambique, Botswana not to mention our perennial customers; the Congolese. Indeed, one-day God will punish this country for not taking advantage of what it has been given. The country has slept for too long and allowed its neighbors like South Africa and foreigners like China, UK, German, India and many others to use its resources. There was an opportunity to turn the country inside out in 1991; but greediness and corruption got into the moral fabric of those that were tasked with the responsibility to provide guidance and that has resulted into a horde of street kids and fallen industries. Nonetheless, the hope is not completely lost if we can accept the condition that we can rise from ashes and explore this potential. If we put our acts together, we could be second to South Africa within a space of ten years at most because of the potential and opportunities that are presenting themselves.

Take for instance; which country apart from Zambia has got Victoria Falls in the region? None, it's only Zambia with about a third in Zimbabwe. Which country has got so many rivers and water bodies in the region apart from Congo D.R.? None, it's only Zambia. Which country has got a diverse of minerals and enjoys remarkable peace? It's none other than Zambia! Zambia can turn its comparative advantage into real competitive edge through effectively exploring its resources by engaging in various economic activities. Nonetheless, in Zambia, easy opportunities lie in agriculture because the country has over seventy percent of the people involved in the sector, and agriculture supports over thirty percent of the industries in Zambia. We can create not less than hundred thousand direct jobs in agriculture per year if leadership puts its policies right and over ten thousand indirect jobs. Giving a practical example, taking a look at operations that has been put up to establish a palm industry in Mpika, once fully operational the enterprise will create thousands of job opportunities. The palm plantation investment will create not less than a thousand jobs at its peak operation. That is a perfect example of investments the country could have concentrated on if it has to be serious will creation of jobs and real wealth for the country.

The activities at Kafue Sugar by another local company in Nampundwe area has created so many job opportunities too. There are not less than two thousand direct jobs created through that investment. If the country had such deliberate investments in each province because the potential is there, so many jobs could be created in a very short period of time. I know that if this government goes on to implement the plans to create a sugar plantation in Kawambwa, many jobs and wealth will be created. This will also prevent people from living rural areas to go and seek livelihoods in Lusaka and the Copperbelt. Already some company is establishing a sugar plantation in Milenge and when fully operational, that investment too, will create several hundred jobs. The fish enterprise through Yalelo and Lake Harvest Hatcheries are such good examples of diversification that have also added to the job statistics for the country. The only thing we need to believe in is that we can do it ourselves. Yes, the country needs foreign direct investment (FDI), but that is not the ultimate. More job opportunities can be created by prudently mobilizing local resources supplemented with FDI. We all need to believe and be assured that an investment in agriculture, even though the benefits don't accrue just there and then, is a worthwhile investment. For now, the country can deliberately ignore the quality of jobs but once the companies are running profitably and well established then it can look at how much they are paying their employees and what value they are adding to the economy. So many economists and optimists have predicted that the world will have problems to feed itself by 2030 in fifteen years' time. There have been increases in energy demands and some farmers are converting their food crops to energy generation which in my view is not a good idea because there are so many alternative sources of renewable energy such as solar that can be explored. Zambia shouldn't have such problems of how to generate energy. If the country can put more investments in the sector, because of so much water due to unexplored waterfalls as well as plenty of sunshine, so are agriculture waste such as cow dung to use in bio-digesters to get methane gas for energy, there will be so much energy generated. There is no reason whatsoever to start converting soybean into energy generation for the industries. This is a unique country which is the only of its kind in Africa but it needs prudent management. No country is like Zambia; therefore, it just needs to work to develop its uniqueness and have everyone enjoying a decent life. Honestly, there is no way South Korea can be more developed than Zambia if the latter puts its acts together! Its people needs to start to refuse some of these shortcomings that befall our country because of lack of proper planning, implementation and political will.

Agriculture is one of the sectors that can create reliable jobs in the interim besides tourism, construction, energy and the manufacturing industries in Zambia. In September 2012, a consultative meeting was held in one of the provinces to brainstorm on some constraints that smallholder farmers face in producing of various commodities. I was taken aback by some of the issues that came out of that meeting and I still can't believe that a country like Zambia who literally depends on mining and agriculture has such challenges in the main sectors which are our cash cows. To be specific, we were discussing the main issues in groundnuts and soybean value chains as regards production and value addition. It's not a secret that Eastern Province is the leading producer of groundnuts in the country. The varieties that are mostly grown are MGV4, MGV5, Chishango and Chalimbana, but amongst these varieties none of them is commercially produced and sold by the seven major seed companies that we have in the country. Because none of them commercially produce the seed, the farmers have resorted to recycling the seed for ages. One would argue that groundnut seeds are not like maize seed which you must buy hybrid seed every year, that is, they are open polinated. However, is this recycled seed treated with fungicides to prevent prevalence of diseases that may come with its production? Not at all! We are talking about diversifying production in the agricultural sector from maize to other crops. The country needs to seriously think about investing into seed multiplication for other commercially viable crops like groundnuts and sunflower. For instance, by only treating groundnut seed with an inoculant, there can be reduced nitrogen fertilizer requirement for the crop by about a third. The country is lucky that some multinational companies such as BASF are promoting such quality technologies to the farmers. They are calling it Histick which improves the productivity of the crop.

Someone may be surprised to learn that the country did not have enough soybean seed in 2011/12 growing season. Soybean was a hot cake and a lot of farmers grew it using grain they had kept in their homesteads. As of November 2012, I had been to the major seed suppliers' depots in Eastern province and never had I found a single bag of soybeans in their warehouses. We need to look at the affordability of seed both in terms of availability, accessibility and the cost. Maize which has received a lot of attention has pack sizes ranging from 1kg up to 25kg. I don't think the commercial seed suppliers package it in

1kgs to sell to big estates like Chimsoro Farms or Mpongwe Farms; the target is the smallholder producer in Nyimba, Shang'ombo, Mpika and many districts in the country. The commercial rate of planting one hectare for maize varieties is twenty to twenty-five kilograms. The commercial rate for planting soybeans is between eighty and 120kg per hectare. The smallest pack size for soybeans is twenty-five kilograms which can plant a Lima or quarter a hectare, while to plant a Lima with maize, one needs about five kilograms of seed. Why then are the seed companies not packaging soybean seed in smaller packs as well to carter for the resource poor farmers that wants to plant less than a hectare? I know we are promoting commercialization but there is always a start for everything. We cannot run away from the fact that over eighty-five per cent of the farmers in Zambia fall in the smallholder bracket, most of who are resource poor and have no access to financing from commercial banks as opposed to the commercial farmers. I know this will defeat the purpose of commercialization of soybeans. However, the initial packaging of seed in smaller packs will be a strategy to promote the production of the crop. These smallholder farmers produce over eighty per cent of the maize that is consumed and traded in Zambia. Why then do seed companies want to continue segregating them from the production of commercial crops like soybeans? It is not a secret that smallholder farmers were not producing soybeans because they didn't have a market where to sell their produce way back, but now that processors want it; they are ready and have come on board. The seed input suppliers are a let down in this regard because most of them have products packaged for commercial farmers. Smallholder farmers are a large market for agro inputs in this country because of the sheer numbers. There are companies in this country that have thrived on smallholder farmers and grown through thoughtful segregation and targeting of smallholder farmers. The input suppliers, therefore, needs to package the products especially inputs for soybeans, groundnuts and sunflower to target this market segment if job creation is anything to go-by. This is a great business opportunity which is lying untapped in the sector. Lucky enough as though they were reading my mind, a company has been established in Chipata which is working with smallholder farmers to multiply groundnut seed. It is called Good Nature Agro and the investment is assured of reaping the benefits because of the demand for groundnut seed in the industry.

If farmers and in particular smallholder farmers are kept marginalized in terms of commercialization/diversification, then the potential for the country will not be realized because they are in the majority in the sector. It is disheartening to note that the private sector strongly promotes small and medium enterprises (SMEs) for economic development in other sectors like trade and commerce, while neglecting or exploiting those in the agricultural production. With the smallholder farmers, labour is the biggest cost and it contributes to over sixty percent of their costs. As soon as they are introduced to modern technologies such as the use of herbicides, efficient yet effective tools for harvesting, they will cut their labour costs by over thirty percent and improve their efficiency. Certain big economies of countries in the world have been developed by harnessing the agricultural sector complimented with other sectors like energy, education and health. It is time the country rightly invested in smallholder farmers through technology and other enablers to create jobs for our country which ultimately will lead to economic development. Additionally, we should invest in value addition to get the most benefit from the industry. One of the mistake that we make is to assume that investments have to be foreign; local investment is more sustainable and last longer than foreign.

Sustaining jobs

Creating jobs through agribusiness activities is one thing and sustaining them is another. As a country, we have been on a drive to create jobs for the ever-growing population which is mostly comprised of the youth. There has been a campaign to make people buy local products. Not very long time ago we saw adverts hitting the media compelling Zambians to buy local products. I will not be wrong to guess that this was after they saw similar campaigns from other countries like South Africa. We saw prominent people in society such as 'uncle K' (not real name) being featured in some adverts on our national television as well as some other legends. However, this drive has of late died a natural death because we no longer see such. Not sustaining such promotions might make people revert to buying South African products that have flooded the market. No doubts that the buy local campaign has had impact on the people's buying behavior. I personally was convinced and saw the need to support the local industries for them to thrive. We all know that developing a product is one thing and being bought is another. If a product is being bought, that is the only assurance that the job(s) that have been created will be sustained. Once a product is successful on the local market, it will be easy for it to penetrate the export market too.

During the first republic, so many jobs were created when local industries such as Mansa batteries were functional but as soon as people stopped buying Spark batteries in preference for cheap foreign batteries from China, all the jobs created went into the drain with the company. However, no one will buy at the expense of compromising on quality and affordability. In 2012/13, I worked quite closely with women development associations in Katete and Chipata as they endeavored to create jobs and wealth through pressing cooking oil from groundnuts and sunflower. This was a way of creating markets for the local farmers as well as improving the nutrition status of the community in which they were operating in. These women associations press oil and nicely package it into various containers which are sold in their communities. I have made it a habit to support these 'cluster' industries and every time I drive to the east (Chipata), I would buy about ten drums of five liters containers which will lasts me close to a year. I have been doing this since 2013. The other thing I have vowed never to buy in some of these chain stores are potatoes. In December 2015, I was surprised to find that a ten-kilogram paunch was costing K80 in one of the famous shops. I vowed never to buy those potatoes and what I do is that every time I drive up north, I buy about two to three pockets at ZNS Naambe in Kabwe from traders along the road. The roadside vendors there buy these potatoes from some farmer on the east side of the road. I know that for every pocket of potatoes I buy, I am sustaining atleast two jobs;

the farm laborer involved in production and that of the vendor selling. If I have no apparent trip to Kabwe, I drive to Chilanga and buy from around that area as there are some farmers growing for Buyabamba as well.

Mealie meal too, is another commodity that are derived from raw materials that are hundred percent grown in Zambia, but I know that I can still contribute to creating of auxiliary jobs of millers. I have taken it upon myself that every year l cultivates a 'shamba' (small field) whose maize I take to the grinding mill for my family to feed on. Similar, a lot of rice is imported in the country although we have our own farmers involved in growing of the crop. From 2007 to 2011, I worked closely with rice farmers and processors in Luapula and Western provinces respectively to help them improve productivity. I have made it a policy that I will be feeding on Mongu, Chiengi or Kaputa rice as long as it is available on the market. I had worked with those farmers before, so I try to support their efforts so that they don't go out of business. Looking smart is something we all thrive to achieve. When it comes to my laundry, no washing paste or powder beats my Boom from Trade Kings. I very well know that Boom is manufactured by a local company; bred and owned by Zambians.

However, I should make mention here that I have sometimes been disappointed by the quality of some of the local products which has forced me to buy foreign products. For instance, I love my relish with hot chili sauce and I have been a fan of Cheek Chili sauce. I used to think it was a Zambian product not until I read on the label that it was a product from Zimbabwe. From that time, I have searched for an alternative product to substitute the cheek chili. I came across a similar looking product which I have chosen not to mention its name here because of what you are about to read. I first bought this product in one of the local shops near my homestead but I was disappointed with the quality. Firstly, though it was written on the label that it was hot chili, the product was mild. Secondly, it smelt of too much 'raw' tomato and the texture of the paste was not as smooth as Cheek chili. I thought that probably it had lost the original quality because of the poor storage in the local shops in the compounds. After about a month of torturing myself to ensure I empty the content, I decided to go to one of the chain stores at one shopping mall and bought a similar product. I was taken aback that the quality was similar to the earlier product I got. I was left with no option but to get the contact details of the company behind that particular product. I sent them an email expressing my displeasure with the quality of their product. I am happy to report that they responded and assured me that it was just one batch which was not produced to specifications. Though I expected them to go a bit further and indicate that they were recalling all products from that batch but they assured me that they will improve on the quality. As we drive the campaign to buy local products, the onus is on the manufacturers to up the quality to match the international standards if they are to continue to remain in business. As Zambians, we always talk about the enormous markets in the region but forget that we are part of the market. We can only be sure of other people buying our products if we are buying them and giving feedback so that they are improved on. You cannot compel a neighbor to eat something you cannot eat as the owner who has manufactured it. The campaign to buy local products should be enhanced in order to create and sustain jobs in the agriculture industry. The potential for job creation is so huge in this sector. So, as a people how are we going to unlock this huge potential for job creation that we have? Let us answer this question.

Unlocking the agribusiness potential

It was a privilege and rare honor to have attended the first ever Agribusiness congress in the country organized by the Zambia National Farmers Union (ZNFU). It was held in Lusaka at InterContinental Hotel in 2012. Congratulations to our farmer union for successfully hosting that mind poking and so informative workshop. People came from all walks of life in Southern Africa and various very important topics were extensively discussed. The only thing that was lacking is that it was not well publicized locally; most of the people only knew of it when they read an advert in their magazine. As Zambians, generally we have a poor reading culture and one would have wished that the advert should have been put on ZNBC, the national television. It was reemphasized how blessed this country is and its strategic position in the region. One of the key points amongst many that resonated so well with participants at the that mammoth congress was that no matter how much money the government will pump into subsidies' the country's agribusiness development will maintain the same level if the private sector like the banks, manufacturing industries and so on don't come on board. This is very true as the government's role is to create an enabling environment for the private sector to thrive.

The truth is that our banks have been a very big disappointment in developing products that favors our agricultural business development in the country. The government gives trillions in kwacha to encourage production but our smallholder farmers have been locked up in the food insecurity doldrums. Although one of the reasons is that the farmer's productivity is too low, financing also has played a role. Our farmers especially the smallholders are asset poor such that they are forced to sell their maize at prices perceived to be too low because of their low productivity. From the same revenue realised they have an obligation to send their children to school and use it to pay other bills. One can only imagine if we had a strong warehousing receipting system in place and how it would have helped them. The poor farmers would have been depositing their maize at the nearest warehouse and go to the bank to redeem/get some money using his receipts whilst waiting for his grain to be sold when the price deemed favourable to him.

From my basic understanding of this system and a bit of my financial information got from the grapevine, no commercial bank is willing to try this system. What is so annoying is that the same banks are implementing it in other countries in the region with less agribusiness potential. Having traversed this region, I have seen this facility being implemented in countries such as Malawi, Zimbabwe and South Africa. As long as our smallholder farmers keep selling their produce at 'low' prices, the benefactors will keep being the traders. As if this was not enough, our public extension services delivery is also not so effective to help the farmers improve productivity. The country has trained so many agriculturalists from many agricultural institutions and other training institutions but the public extension system is the most ineffective one can find in the region. It is not that they don't know what they are supposed to be doing but they are poorly funded if ever funded at all. Their funding is inversely proportion to what they can do.

However, some of the officers have even lost it in that they have developed a poor work culture of 'I don't care' as long as their bank accounts are debited at the end of the month. These smaller private companies that are involved in selling of inputs such as agrochemicals, fertilisers and seed have a more effective extension system than our public one. The officers are not provided with adequate tools to work or effectively carry out their duties. If only the country could develop its extension system by effectively providing the tools for work to extension officers including retraining and having an effective monitoring system, the farmers' productivity will improve. The government too, through the bank of Zambia should develop financial policies which will encourage banks to lend money to SMEs and individual farmers. Our banks have concentrated on lending to the government and civil servants because they perceive them to be less risky than farmers. Lack of agro-financing in the sector has also contributed our farmers to remain small scale for ages; some even feel good to be called by that name even if they are cultivating more than ten hectares. By definition, in other countries, a farmer cultivating ten hectares is considered to be commercial in other countries because they have adopted production of commercial cash crops. For instance, a farmer cultivating ten hectares of tobacco in Malawi is way better off than a farmer that is cultivating fifty hectares of maize in Zambia.

Lastly, many of the people's understanding of agriculture is limited to one crop. Agriculture is broader than maize growing *inobantu* (you people). This is because we have kept developing policies that are centred around maize for a very long time. If we keep developing policies that are centred on improving the maize efficiency, then we should forget about unlocking the agribusiness potential that Zambia has. We have emphasized so much on maize such that our public extension officers find it difficult to even advise a farmer on how to grow other crops such as sweet potatoes. When we go for traditional ceremonies like *umutomboko, ubwilile, Nc'wala*; we see a lot of crops such as millet, sorghum, *busala, chinangwa* (cassava), *intungulu* (wild fruits), name them; but none of that is in our policies except where they are mentioned in general as a way of diversifying. We therefore need to allow the farmers chose commodities that will make them grow or graduate from the unfortunate name tag of small scale farmer. Partly, the government is to blame because they have made the farmers to believe that one can only be called a farmer they are cultivating maize by limiting the support to maize. This has even been twisted with coining of terminologies such as food security; which in the mind of households mean having adequate maize to feed in a year. I am looking forward to a day when we are going to have a leadership that will take to task farmers that will only be cultivating maize. I feel that we have a ripe environment as discussed earlier to develop our agribusiness sector in this country; we all need to share in that vision. A few things can be identified that point to the fact that Zambia's agribusiness environment is ripe and moving in the right direction by some recent happenings in the industry.

The ripe agribusiness environment

For those that could have been in Zambia in 2004, how many multinational agribusinesses did we have in the country? There were very few if any but to date there are so many that have opened operations in Zambia. There are those in the input supply such as Monsanto, DuPont, Pannar, BASF, Bayer, Syngenta, Aryster, Omnia, Yara, Amiran, ETG while in the off-take markets are NWK, AFGRI, Mt Meru, LDC and many others too numerous to itemize. Reports reaching my desk are that another multinational fertilizer company will open operations very soon (it had already by the time of publishing the book). The country is also blessed with companies that left the country at the time it had challenges who are coming back to set base again. Just the other day, I was perplexed to see a Dunlop branded company driving down our streets in Lusaka. At the time I was growing up, tyres were synonymous to Dunlop; they are back too!

This is a good sign that these companies have seen something good in Zambia's (agri)business environment despite having a weak and unstable currency. However, what still needs to be done is setting up operational factories in the country; with an exception of Mt Meru, Cargill and to some extent Greenbelt many of these companies are yet to establish processing units or manufacturing base in Zambia. It is only the seed companies that have fully intergrated operations of seed business done locally. This still gives me a nagging headache because tomorrow they can easily decide to close shop and leave. I think the government through its commerce minister should revisit their investment policies. Just imagine, Kenya, South Africa and Tanzania have some plants for some of these companies; for instance, we know that many of the agrochemical companies have plants where they manufacture and package these products in Kenya and South Africa. Why should our South African brothers be packaging these products for us when we also have the potential and labour to do that? The country need to have such plants established locally so that the job opportunities are enhanced and products will become competitive. The excuse that Zambia has no port for easy shipment of raw materials is neither here nor there because the country can now have access to Walvis Bay which is a stone through away from Livingstone. Raw materials can easily be brought through that port and have them repackaged here in Zambia. The country can't just be a surplus producer of grain in the region when it is not benefiting on the least cost of inputs. There is need to have fertilizer packaging projects just like those found in the region. Zimbabwe did it before that country embarked on the land reform program, most of the multinationals had plants and offices there. The advantage in Zambia is that it has been a peaceful nation with relatively stable policies since independence. However, at the time of publishing the book multinational companies such as BASF, BAYER, Syngenta, Yara, Omnia and many others had legal entities with offices in the country.

Zambia has the right agribusiness environment for investments in the region. Not long ago most companies used to rush to South Africa but they have just realized that South Africa's potential to improve productivity is almost reaching the upper limit. In Zambia, there is availability of land, with low productivity, it is an opportunity for investment in various technologies to improve productivity. Even with the low productivity, the country has managed to feed the region in many years and it is still producing more maize, soybeans and groundnuts even during very drier seasons than normal. This shows that there is a comparative advantage over the neighboring countries in the region. This country is amongst the few countries that have invested massively in infrastructure development; there are roads being built everywhere in the country. This is bearing fruit as can be seen by multinational investments in rural areas. Local companies are also coming on board such as the milling plant established in Chipata and the processing plant by COMACO in the same town. Though we are not able to realise the economic benefits from some roads investments now but they will add value to decentralization and opening up of remote areas for other investments and development. I was last in Mansa in 2009 but I was elated to learn that there is a sugar plantation being established in Milenge district. So is Shangombo in Western province. My earnest appeal is that the government comes up with policies which will compel commodity traders, for instance, not to be exporting raw materials like maize that was recently impounded in Chipata destined for Malawi. They should allow exporters to export either mealie-meal, animal feed or *Junta* (local beer) instead of maize grain. This has to start with government; we are aware that our colleagues from DR Congo sent envoys to president Lungu to ask for help in terms of food. The agreement should be that we can help them with food in form of mealie-meal and not maize grain. We are seeing massive investments in power generation and distribution. The Kafue gorge project is having the capacity increased; the Maamba thermal plant has just added 300MW of power to the national grid; the Itezhi Tezhi is another one under renovations as well as the Musonda falls in Luapula. Power is very important for the manufacturing industry. As discussed earlier, there have been massive improvement in the transport and communication sectors though much still needs to be done.

The plea of Zambians is that all those multinationals that are citing on the fence should immediately start thinking of opening operations in Zambia. We hope to see Zambia become the Dubai of Southern Africa. The backbone of Zambia's commerce industry lies with agricultural development with supportive sectors such as energy, transport and communication as well as tourism. One area Zambia is endowed with is availability of fresh water. We have so many rivers and lakes in the country though only one percent of Zambia's irrigable land is under irrigation. For instance, with the strides in agriculture there is still much to be done in irrigation to improve on the water utilization in the sector. Great opportunities lie in developing the irrigation in the agriculture industry.

Irrigation development in Zambia

During the time I was writing this article, it was the seventh day in the month of January in 2015 and yet the country had not had a serious onset of the rains in Lusaka. In olden days, this is the time when the country could have been receiving rains almost every day and sometimes not even seeing the sun. This is really disturbing to the farming community who enjoys seeing the green color. It is not a secret that we have a drought as predicted by weather pundits who warned us that we were going to have an El Nino weather pattern in this part of the world.

For people that might not know what the implication of El Nino to this part of the world is that it brings very dry weather conditions. It must be given a local name which people can easily associate with. 2015 was really a bad year for Zambia as far as agriculture is concerned. However, during a crisis, it is the best time to look at hidden opportunities and explore them. Always when there is a crisis, lays so many opportunities as they say every coin has two sides. What Zambia can do to improve on its agriculture lies with stride in developing an effective irrigation system. Well, it is well-known that Israel is drier than Zambia but that country is able to feed itself because of the technological advancements in irrigation. Most of Egypt is a desert and they only rely on one river; the Nile River and it is annoying to observe that some of the products sold in the supermarkets in Zambia such as juices and other foods are imported from that country. Agriculture now should not depend on rainfed.

Indeed, even irrigation to some extent depends on how much of the rains we receive that particular year to fill up the dams. In Mkushi farming block, farmers in 2015 cultivated only about forty percent of what they normally do in winter because most of the dams were empty. Even if dams are built, if the country does not receive adequate rains in a particular season, there will still be challenges with irrigation. Therefore, what could we have done to effectively develop this irrigation system? Some newspaper in that particular year made mention that there are some documents that are accumulating dust in some government office where it was decided that water could be transported through canals from the northern parts of the country to the south for instance. The paper quoted one of the prominent citizens of this country who at one time served as the republican vice president as having being the source of that valuable information.

That statement made me think outside the box, and at first I thought it would be an expensive venture, but nothing will ever be done cheaply because we have destroyed the good environment God gave us through industrialization and we need to pay dearly. It is unfortunate that countries that contribute and have contributed too much dangerous gases which have eaten away the ozone layer are siting on the 'fence' to contribute to its remedy. Most of our river sources are in the north near D.R. Congo and they run southwards to offload their waters in the mighty Zambezi River. Every year the country lets trillions of liters or gallons of water to run into the salty Indian Ocean. If there were heavy investments, each of these rivers can be linked to the other because they run parallel to each other in most cases. Possible points in the channel of each river can be explored such that canals can be made and underground channels where there are uphill gradients for instance, to link Kafue River to Zambezi River. Along those canals, several dams can be developed and farmers along those canals can also draw water for irrigation. The same can be done with Luangwa River and the other streams/rivers that run parallel to it. The reason being that even if we build dams along the perennial streams, if the country does not receive adequate rainfall like was the case in 2015, those dams to be developed will be white elephants. The government is putting up irrigation schemes along Momboshi river in Chisamba and somewhere in Mufulira but with conditions like was the case in 2015, there are huge doubts whether the Momboshi irrigation scheme will be operationalized and put to good use unless wonders happens. The country has to find ways of letting water remain longer inland before it gets to the Zambezi River beyond the Kariba dam. Though these are ideas of a layman and not a water or civil engineer but I guess the country needs to start working on feasibility studies of making this happen. There is abundant water in lake Bangweulu which does not dry and engineers could find ways of building canals from Luapula River after the Tuta Bridge to feed into streams and rivers that feeds into the Mkushi farm block. These are massive projects which need great investments but this cannot be done in a year. This might sound weird now but definitely, the country will one day need such projects to be done in Zambia. There should be great thinking about this and share information where possible.

The potential for irrigation in Zambia is so immense and once developed, it could easily contribute to productivity improvement. By improving the irrigation prospects in agriculture, practicing farming as a whole year-round activity as well as introducing new commodities in our farming systems would be a possibility and one can only imagine what that could do to the economy. Irrigation is indeed an activity which every farmer that practices farming as a business should embrace. If we start using irrigation on our crops, we will be nearing to getting the actual potential of every crop and this will lead to achieving bumper harvest that we normally talk about. What is a bumper harvest and has Zambia ever recorded a bumper harvest before?

Planting a 'bumper crop'

Planning is a very important ingredient for any success in business. There has never been anyone who has succeeded in life, business, and sport, name them without a thorough planning. You will agree with me that even national plunderers in many countries plan how they will steal public resources. Many plunderers have been studying the government systems for years before they ventured into their evil and criminal acts. The point I am trying to drive home is that planning is the cornerstone of success. The Formula one road project that was embarked on by the previous government failed (to be completed in time) because it was done without strategic planning, but Kariba Dam has stood to this day even though it was built in the 1950s because there was planning involved before it was built. We have electricity shortages today because people tasked with the role of planning to expand the infrastructure went to sleep – deep slumber for that matter!

You might be wondering as why I am belaboring to elaborate about planning when am supposed to be talking about the trillions that was paid to maize farmers by FRA in 2012 when the budget for the ministry of agriculture was less than what was paid. How possible is it that a department of a company can make more money in a fiscal year than the company itself (only with FRA)? I presume that many of the farmers that cried that FRA was not paying them in time did not plan well. Why should one continue to deal in business for a commodity when year-in year-out, they are faced with the same challenges of late payments? Farmers may give an excuse that they grow maize because it is a staple crop and it has a ready market. They could be right in that the commodity indeed has a ready market with a premium price of K75 per 50kg bag ($220/Mt) if sold to FRA. But do these farmers know that they can get more benefits if they grew onions, for instance instead of maize? Let us go through a simple example which might stimulate our thoughts of what many of us will plant this season. In Zambia, vegetables that includes tomatoes, onions, cabbage, rape and many others are grown in the dry season from April to around November. Immediately, the heavens open (rainy season), most of the farmers turn to growing maize and other field crops such as groundnuts, sweet potatoes, soybeans and many others.

Do farmers know that in some instances, this country imports onions from South Africa especially during the period February to April of each year? The gross margins for onions in most cases is not less the K5,000 per hectare as compared to maize gross margins which is around K300 and negative in most instances for our smallholder farmers. The average maize yield per hectare for most of the smallholder farmers is around 2.2Mt/ha. During the period, I have mentioned, we have had certain commercial farmers that produce onions using intensive agriculture technologies in Lusaka, then offload the onions on the market, and during that period one bulb is costing around K0.50. Suppose that a farmer cultivates three hectares of maize and gets 6.6Mt (132 x 50bags) of maize. The total income the farmer will get in November for instance, when FRA pays the farmers would be K8, 580 if everything was sold to FRA. On the other hand, if you decide to grow one hectare of maize for your own consumption from which you will get 44 x 50kg bags of maize which should be more than enough to feed an average family of six people for the whole year. With the remaining two hectares, you could plant one hectare of soybeans from which you can harvest 20 x 50kg bags (at 1Mt/ha yield rate), and the remaining one hectare is divided into two halves; one half is planted with groundnuts on which you can get a yield of 0.5Mt/ha (5 x 50kg since its half a hectare) and the final half hectare is planted with either tomato or onions. The yield potential or actual yields that farmers get from onions and tomatoes is not less than 20Mt/ha.

If the tabulations done for the second option, you will not get less than K15, 000 of net incomes. Does this compare with the K8, 580 (gross) which you will only be able to access in November? I do not think so! Besides, the groundnuts can be used to improve the nutritional need for your household, so does the soybeans while you will have money throughout the year because these commodities/crops mature at different times. You will also agree with me that option two avails you an opportunity to practice crop rotation as well as crop diversification. In the third quarter of 2013, I received so many emails requesting me to guide some farmers of what crop they need to grow that rainy season. I should state here that in most cases, I have not told anyone what crop they should grow because;

i. I did not know the areas where the farmers practice their agriculture from,
ii. I did not know what inputs/resources is at their disposal, and
iii. I don't know how good they are in terms of agronomic know-how about the crops I can advise them on and finally the market prospects of different crops in their areas.

For those, wanting to go into horticultural production for instance, I had referred them to friends that are in the input supply near them such as those working for BASF, Agrifocus, Syngenta, Bayer, Amiran and other such companies.

For anyone that might want to get advice on the specific crop they want to grow, they need to seek advice from some input suppliers, ministry of agriculture extension staff or lead farmers in the area near where they operate from. Additionally, I had been writing about market opportunities for several commodities since 2012 to May 2016. It is the responsibility of a farmer as a businessperson to conduct a market research and assess which commodity could be fetching better prices or profitably on the market. Time of growing a crop just for the sake of growing it is long gone. In addition, let me emphasis that farmers should develop a habit to have their soils tested from their fields for various elements. It is not as expensive as they think it is because a sample for analysis of N, P, K, Ca, Mg, Cu, Fe, Zn, pH, and S costs between K180 to K250 with most of these learning and research institutions. This information can be of very important about what fertilizers to use, and this can make a difference of getting a bumper harvest and not. You may not get a good harvest with a blanket application of fertilizers as per recommendations from 1960s. Soils do vary within a field as well as the country. The amount of fertilizer and type that you may apply in Mpongwe will differ with that for Mkushi. There are issues to do with soil pH, texture, organic matter and many other factors that we take for granted in our farming enterprises.

To generally answer to some requests, I have been getting from farmers asking about what maize varieties they can plant, let me further write on what you can do to get your good yields. Some have further asked whether they can get two hundred bags of maize in a hectare. Others have asked about where they can get seed for soybeans as well as the HiStick inoculant. For me, this is a good sign that farmers are learning to plan their businesses unlike what used to happen where one will grow a crop just because the neighbors are growing it. Having gone around in the region; East Africa, Southern Africa and a few countries in West Africa, our smallholder farming is getting somewhere. My only worry is our partial adoption of production technologies, where you find some farmers are only using seed and fertilisers. Most farmers are now using certified seed but very few are applying herbicides for their weed control and none is using fungicides for disease control in crops like maize and later on insecticides for pests except for the dreaded fall army worm. Farmers were caught napping for instance in 2016 when there was an outbreak of Fall Army worms. Those that practice preventative control had a very good control of this pest while others that do not had a shock of their life time because this is a very difficult pest to control and wiped out some fields reducing the yields drastically.

This is so worrying for me; it's like having a child and expecting that child to reach adulthood without having malaria (falling sick) when they are being bitten by mosquitoes. In the villages and certain communities, they will also have to cope with having lice (*indaa*) in their hair. This is part of life and growing, crops too are living things which needs 'medicines' to fight 'illnesses'. Well, let me state here that tomato and vegetable farmers have learnt this art because they cannot do away without fungicides, insecticides and nematicides in tomato growing; back to the question, are you able to get two hundred bags in a hectare? My answer is a big yes, and some smallholder farmers are getting that and more. Mind you, it is only by improving your productivity that your agribusiness will become competitive and profitable. Without going into calculations, one is only able to make a profit if they are producing above five metric tons for maize, serve for this season which has been special because of the overwhelming demand for the commodity in the region. Probably the question we need to be asking ourselves is what should we do to get two hundred bags and above! Good clever question from a thrift *agripreneur*!

First things first, choose the right variety of seed that suits your agro-climatic region in which you are practicing your farming business. You know by now that there are basically three types of maize seed on the market in Zambia; early maturing for regions that receive up to 400mm of rains, medium maturing for those receiving up to 800mm of rains and the late maturing for those that receive over a thousand. However, it does not mean to imply that the said varieties cannot grow in other regions, they can but you either have to supplement with irrigation or need to time the planting dates for instance, in the case of early maturing in high rainfall regions. Once this is done, the next thing to do is control or manage the weeds. I know farmers especially the smallholder would want to apply herbicides after seeing the weeds germinate in their fields. There is a perfect product that you can use; it is called Stellar star! This is an innovation which will only select and kill the weeds while leaving the maize standing. The science behind this action is beyond the scope of this book because it is too scientific to understand for my grandpa in the village. You again must time your application; don't allow your weeds to grow very big but apply while weeds are not taller than your middle finger or five centimetres. Timing in weed management is everything and this is what separates good farmers from average ones. Nonetheless, for those that can afford, apply the pre-emergence herbicides such as Integrity and Lumax which just like stellar star will control both broad leave and grass weeds.

Once this is done you are on the right path to getting your ten tons which is the average yield for our friends in USA. The other ingredient is the fertilizer application. Through the rule of the thumb in Zambia, many farmers are using 200kg of compound fertilizer like D-Compound and same amount of urea or ammonium nitrate. D-compound fertilizer should be applied at planting or just after the maize has emerged. Common mistake made is that you wait after you have done your weeding and if you are using a hoe, this might be after four to six weeks after planting – this is too late! If you did your soil analysis, you might need less of the fertilizer or more. It's like having a child and wait to breast feed him until he starts laughing; he will not survive. The amount of fertilizer to apply especially the top dressing may also depend on the plant population; the higher the plant population the more fertilizer you will need to apply. Lastly, maize is prone to many diseases especially the fungal diseases. Some are very difficult to know whether it's a disease unless you have an expert eye but the prudent thing to do is to apply these pesticides ('medicines') even before you see the disease, and this process is called preventative application. One of the best products I have used and with very good results is dear "Opera" and Amistar Top, lovely products which if you apply twice in a season, it will not only protect your crop from diseases but will give you an extra leg in stress management of the crop and quality of the produce while enhancing yield above the normal products as high as thirty percent. For the Opera, it has a greening effect property which is called AgCelence by the manufacturers of the product. There is no way you will fail to get your two hundred bags if you follow this program judiciously.

Indeed, sometimes you may be wondering as to who steals your nutrients which you apply in your fields. The battle to get the good yields (220 bags) of maize grain starts a few months just after harvesting. Farmers that are serious with getting good yields have all their inputs ready and fields prepared before the onset of the rains. Many of them are now planting their maize with the first true rains. Remember in business you are always planning and reviewing your plans. However, a few of some farmers don't even know the variety that they will plant as late as November when the season has set in - that is suicidal! Such farmers are not businesspersons but gamblers.

Remember nurturing a good crop is like taking care of your lovely child; the care does not start when the child is born but at the time you are planning to conceive. Many families will start buying clothes such as nappies as soon as they know they are pregnant. If you are a father that normally likes socializing, as soon as you know your wife will be a mother you even reduce or stop your 'throat' activities to take care of the new family member. It is like farming; I know of a close family member who stops drinking the first day of October until May when his harvest is ready. Many of us farmers do not want to even go to the field to admire our maize as they 'wave' to us with that golden green color; the color I treasure most. Therefore, what are some of the things that steal nutrients from our crop which we need to be aware of?

The first thing you need to be aware of is that you need to have a crop with healthy root system; this is the foundation for effective crop growth. To have such, you need to prepare your land so well, it should not have a hard pan. This can be done by regularly ripping your field at least once every three cropping seasons. Secondly, you need to dress your seed with some seed dressing. Lucky enough, for most of the seed sold on the market it already comes dressed. However, if you can afford, there is no harm in dressing it again because you don't know in what condition the seed was kept especially in some of the warehouses of rural agro dealers. Seed dressing is the foundation for protecting your crop from being attacked by soil pests. Thereafter, you need to ensure that the crop finds the 'food' as it germinates. Normally most of us smallholder farmers wait to apply basal fertilizers until after we have done our weeding; that is wrong. It is like waiting to give breast milk to a child until the first milk teeth appears, will it survive? Apply your basal fertilizer at planting so that the germinating seed will find the nutrients in the soil, which is the only way you will have a healthy plant.

Thirdly, there are so many thieves that steal the nutrients from the crop in the soil. One of the great thieves are the weeds. Weeds can cause total crop failure and we have seen this with a lot of farmers that have abandoned their fields because they have realized that they will not get anything from their fields. As I have always said the best weed control is at pre-emergence; that is before the weeds and the crop emerges from the soil. Using post emergence herbicides is normally a fire fighting activity though most smallholder farmers prefer it. However, post emergent if done when the weeds are still young can also be effective. Weeds steal about half to over eighty percent of the applied nutrients. Worse off, there are certain weeds which even lodges its roots directly in the plant roots and gets the nutrients like tape worms, these are the ones that we need to target at pre-emergence. Fourthly, there are so many insects in the soil which feeds on the fertilizers or nutrients. I would not like to go in the soil chemistry of how the fertilizer is taken up by the plant as it might be confusing to you, but what you need to know is that there are two types of soil microorganisms found in the soil. One is beneficial to the crop as its breaks the fertilizer in elements that are easily taken up by the crop. On the other hand, there are other harmful microorganisms that steal from our crop at root interface; examples of such insects are nematodes. These must be gotten rid of by application of nematicides.

However, in Zambia though we have nematodes that affect our maize, we do not have an effective nematicide ready that can be applied to the maize crop. Nonetheless, treating of the seed will help reduce the impact. The last two types of thieves that steal from our crops are pests like aphids and disease causing fungus such as rust and others. These are easily controlled but some farmers have paid a blind eye and that is the reason they have been failing to get 220bags per hectare. Are you going to allow thieves steal from you this season? In the preceding article that was published in the Times of Zambia in May 2012 and it was on tomato production, I will try to show how tomato growing is a profitable enterprise in Zambia and it has made some farmers become millionaires.

Growing tomato is profitable

TOMATO to most of us is a vegetable because of the way we eat it, but botanically speaking, it is a fruit because it has seeds, and it could also be classified as a berry since it is pulpy and has edible seeds. Though we have classified it as a fruit, I do not expect someone to go into an orchard looking for tomatoes to eat them like oranges. The scientific name for tomato is *Lycopersicon esculentum* and it belongs to the same family as egg plants, tobacco and many other solanaceous plants. It is native to western South America but has been domesticated in all the countries in the world through the process of plant breeding.

Even though it has been classified as a fruit, it will be refered to as a vegetable in this article. This fruit is widely grown in Zambia and almost all households use it or at least they eat it twice or more in a day whenever having meals. It is a fruit that made a lot of people leave Lusaka for the country-side, and a lot of them that have small-holdings in peri-urban Lusaka gets their livelihoods from growing this fruit cum vegetable. Many may wonder why they should venture into the tomato business when over forty per cent of farmers in Lusaka for instance, are growing it. Despite the many farmers that are growing tomato, the demand for it is enormous and ever growing. This is because people eat it every day and therefore, the demand for this commodity is an everyday event. The market for this commodity can be segregated according to the quality that is produced.

Most of the farmers who produce this commodity with less inputs, especially the disease and pest control products, makes them fail to supply markets which pays a better premium price like chain stores, hotels and lodges because of the poor quality of their produce. They sell their product on the open market due to quality issues. Therefore, when targeting lucrative markets, quality is one factor which should not be compromised. The timing of when to grow this commodity is also very important if one wants to get the most out of it. For instance, those that grow and have the fruit mature between January and April gets the best returns out their production because the prices are normally higher during that period. Just last week (February 2012), my wife had asked me to get her some tomatoes when coming from work in the evenings. I bought three medium size tomatoes at K3.00, implying that one tomato was costing K1.00. How many tomatoes can one harvest from a well grown plant in a season? The above traders/farmers were making a kill and getting the most benefit, this is because during that time (rainy season), most farmers fail to control the problem of the tomato flowers falling off (flower abortion). This makes the commodity quite scarce during this period of the year unless one is using the green houses or if one has got enormous experience in growing tomato. There are also a lot of diseases during the rainy season such as fungal and bacterial diseases. Hence, the quality of the fruit is also affected.

For instance, one small-scale farmer bought a five-ton Canter truck from the harvest of tomato in 2005/6 season from an area of about a hectare that he was cultivating. During the rainy season of that year, prices of tomatoes soared as high as K120 ($40 then) per box at Kasumbalesa border with Congo and it was as high as K80 ($27) at Soweto Market in Lusaka. This farmer would harvest hundred boxes a day and take them to Kasumbalesa, on his way back; he had K12, 000 ($4 000) in his pocket before deducting the transportation costs which mostly was fuel, levies and handling costs. However even after deducting the transportation costs he would remain with not less than K9, 000. Interesting enough, when everyone was so taken up in maize production due to the Farmer Input Support Program (FISP), this thrift farmer would collect the fertilizer and use it for his tomato enterprise. The other best time when one needs to manage the crop well is during winter; because again, much of the crop is affected by the coldness and very few farmers will successfully manage it by the time they will be harvesting. Remember that tomato does not produce fruits (normal) when temperatures have fallen below ten degrees Celsius (10°C). This means there will be low supply of the commodity on the market making them fetch good prices following the law of demand and supply. With farmers that have adequate resources, they would produce them under greenhouses during these periods to compensate on quality.

Some farmers had been complaining to me that they had decided to go into soybean production and everyone else was now growing it; therefore, they were worried that the prices might go down due to high supply. I assured them that the prices for soybean will never fall below K2.00 per kg (2012). Surprisingly enough, some of those farmers also rear pigs and poultry; they buy the feed from the market. I wondered why they couldn't make their own feed for the piggery and poultry enterprises if they see that the prices were suppressed due to supply. I told them about a friend in Mkushi area that was doing the same and he is making more money than he would if he was to sell it to millers as a commodity. Business of farming just like any other businesses requires that you develop a strategy that you follow through, it is not just all about cultivating.

The year before this article was published during the AGOA summit that was hosted here in Zambia; we heard that FreshPikt[5] a Zambian owned company was going to be supplying the American market with processed beans and tomato paste. This was an opportunity to sign supply contract with them if one couldn't break through the market offered by the chain due to quality or other trade barriers. The chairperson of that company was Zambian and I believed he would proudly support Zambian products as long as they are of good quality. Tomato just like any other crop is also affected by diseases, both fungal and bacterial. You will note that most of the bacterial diseases will come about due to secondary infection. For instance, they enter through 'wounds' left by spider mites or aphids and sometimes physical damage as we are working on the plant such as when staking or pruning

---

5 FreshPikt has since gone under receivership and its assets has been bought off by ETG group of companies

the plant. Therefore, it is important that we take care when working in the field and always use sanitizers. Mostly symptoms of bacterial infections are abnormal growth, rotting, and wilting. The important distinguishing characteristics of a bacterial infection are the sticky and slimy materials secreted by the bacterial cells that later become smelling like fish. It is advisable to use sanitizers such as spore kill when we are pruning the crop.

For our own sake, some of the fungal diseases of economic importance to tomato production are;

Early blight - although called early, it usually sets in late in the growing season and this is the disease which renders the tomato to be watery. Blight is a disease condition wherein an affected plant part is dried or dies and has yellow around the dead tissues. It also causes sudden, severe, and extensive spotting, discoloration, or destruction of leaves, flowers, stems, or the entire plants and usually attacking young growing tissues. It is however, a disease which can be managed with some products currently being sold on the market such as Bellis and Ridomil supplied by BASF and Syngenta respectively. None of these two products should be sprayed more than twice in a growing period of a crop to avoid developing resistance of the disease.

Late blight - it usually sets in earlier than early blight and on leaves, and it can sometimes be confused with nitrogen deficiency. Both the above diseases can be remedied with products such as a combination of Mefenoxam and Mancozeb called Ridomil Gold from Syngenta. BASF, a German chemical company has a good product called Acrobat, which is a combination of Mancozeb and Dimethomorph as active ingredients which is sprayed preventively and has to be sprayed every fortnight. These indeed are very good products that a tomato farmer must always have. There are some other products on the market that can control the two diseases. Under normal circumstances, tomato should be sprayed on a weekly basis with protective fungicides such as Mancozeb or any copper based fungicide such as copper oxychloride or copper hydroxide. These should be alternated on a weekly basis. These two diseases are quite pronounced during the rainy season due to the favourable conditions which favors sporulation such as high humidity and complete cloud cover.

The other fungal diseases that affect tomatoes are *Powdery* and *Downey mildew*. This is not as stubborn as the blights and can easily be controlled with most of the curative fungicides. Mildews are the white spots or patches on leaves, shoots, and other plan parts. *Downy mildew* kills the infected plant fast, while *Powdery mildew* will cause stunted growth, stressed plant, and reduced yield but rarely kills the plant. In the rainy season, one must be on the lookout for many other diseases such as the blossom end rot as well as botrytis. Bellis mentioned above will control Botrytis very effectively, not only in tomato but other crops as well (always read the label).

I have be-labored to go into the details of even prescribing some pesticides (chemicals) that can be used on tomato production because of the quality of tomatoes that I bought on the road side the other week (October 2012) in Katete. The tomatoes were well fed looking at the size of the fruits but they were neglected as they had a lot of spots due to insect damage. The price was low because of this same problem which could have been avoided. Some insects like aphids and whiteflies can be controlled even with organic chemicals such as neem solutions, onion/garlic solutions and many other herbs. However, in tomato production the most devastating pest if it sets in is Red Spider mites. These are very small orange insects which can completely wipe out the crop if not effectively controlled. It is important to regularly scout your fields for these pests and once identified or noticed, don't wait for them to reach a spraying threshold. Products like Abamectin will control the adult insects while Tedion (not too sure whether still on the market) will control the nymph and egg stages. I should confirm that I have for a while not been able to see Tedion on the market but there are other good products on the market as well. The most important thing to do is that when one is planning to venture into such enterprises, it is important to seek expert advice from agronomists that work with seed, fertilizer and crop protection companies that are available in all the countries in Southern Africa including Zambia.

The next time you see tomatoes, remember how they were produced and know that some asset poor farmers have made their livelihoods from growing and selling tomatoes. They have made more money than even some emergent farmers that have several hectares of land under field crops such as maize. Remember that being food secure doesn't mean just having enough maize! Hey, a friend of mine has been making money by keeping fish in ponds. It wouldn't be a bad idea to talk about keeping fish. Fish should not only be eaten in Mongu or Mansa or Kariba, even people in Mgubudu, Nakonde, Choma and Kapiri can eat the famous pale! For now, let's continue to discuss whether tomatoes are difficult to grow because some farmers find it difficult to effectively grow tomatoes.

Are tomatoes difficult to grow?

When the previous article on tomato was published in the newspaper, I received so many questions and commendations from people that found the article so helpful in their tomato production enterprises. Jokingly some went to the extent of wanting to offer me jobs on their farms. I made mention to them that mine was not about getting money from them but to offer the most needed information which might be helpful in growing their various enterprises. The purpose of my writing weekly articles in the paper was to share information about various enterpirses in order to improve productivity. I have this feeling that once productivity for various crops improves, farmers will make money and everyone will be happy including me because I will have a wider base to offer my consultancy services – it is a case of making the 'cake' big before sharing it.

Indeed, tomatoes are not an easy crops to grow and they are not very difficult to grow either; the basic fundamentals about its production are followed. This plant belongs to the *solanaceous* which is a very important family. Potatoes, egg plants, chili and many other crops belong to this family too. They can either be determinate or indeterminate just like some soybean varieties. Indeterminate varieties bear fruit continuously and the opposite is true for determinate. The choice on what to grow will depend on the level of management you want to employ and the experience that one has in growing of the crop. To the new comers like my colleague from German who expressed interest in venturing into the enterprise at her Chongwe farm, it would be better to start with determinate varieties until when you have gained enough experience in tomato production. To produce a perfect tomato fruit, always be alert and keep an eye on your plant's health, because this crop/plant is attacked by an array of diseases and pests. Some farm managers have ended up being fired because they failed to control the diseases that could wipe out the entire crop in the field.

The major diseases to watch out for are the fungal diseases as alluded to earlier such as early blight and late blight. These are diseases which if not managed well can reduce the productivity of the crop or wipe out it entirely. They are more pronounced in the rainy season because of favorable weather conditions that favor sporulation of the fungi spores. Though early blight comes later than late blight in most instances, sometimes it can set in early. The general physical features or symptoms of the diseases are primarily found on the leaves but may also cause the fruit to rot near the stem. Symptoms first appear on older leaves and are characterized by irregular shaped brown sunken spots with concentric rings. The tissues surrounding each spot usually turns yellow and die back before the leaves drop off. On the other hand, late blight is caused by *Phytopthora infestans* which is one of the most notorious and devastating organisms because the *sporangia* or *mycelia* (disease causing organisms) are dispersed from infected plants organ by wind, splashing raindrops or wind driven rain. The leaves and tissue including the fruit will dry off as if there was wild fire. You will agree with me that rain fed tomato is quite difficult to grow than in the dry season unless it is grown under a green house. However, there are less notorious pests during the rainy season than the dry season. The most common pests are the aphids and fruit borers. On the other hand, a lot of people have failed to manage the tomato crop even potatoes in the dry season because of the presence of mites called the red spider mites. These are quite destructive and have the potential to wipe out the entire crop. This does not mean that you will have no red spider mites in the rainy season – far from it! It's just that they are more destructive in the dry season than rain season. The best way to control diseases and pests like I have always emphasized is to regularly scout your crop and institute remedial measures of spraying preventatively unlike firefighting. I remember a combination of products like Tedion and Abamectin combating this problem while fungicides like Amistar, Amistar Top, Revus Top, Bellis, Copper based fungicides and Bravo are very handy on fungal diseases. Whiteflies are the other pests that attack tomatoes and they can reduce the marketable value of the fruit. The other thing that we always want to control pests such as the aphids is their potential to transmit other deadly diseases like viral diseases. Virus are never controlled by any known pesticide and once the crop is affected, the best control method is to rogue out affected plants and destroy them.

The farmer should know that these are not the only diseases and pests that attack your tomato; remember that there are also other pests like nematodes. These micro living organisms are very important to control if you want to grow a quality crop. Unlike the others, these compete for nutrients by infecting (affecting) the roots as opposed to the leaves. Like I always say, the only successful farmer is one that is not shy to ask the experts or the best farmer in their area. It is more difficult to grow tomato than it is to grow maize but I always advise the farmers that you will always make good money from a difficult crop than a simple one. Do you know that you can easily buy a 5Mt Mitsubishi Canter Truck from growing a one-hectare tomato crop but you will **never** buy it with a maize crop! I have shining examples of some of the farmers that I advise that have bought brand new vehicles from the showrooms through the business of growing tomato. Tomato growing is business while maize growing is a way of life or food security unless you are getting over 5Mt per hectare. Onion is game too! If you have any questions and don't keep it to yourself or make assumptions as most farmers do. Find the courage to ask questions and to express what you really feel and want to know from experienced farmers or extension workers. You can completely transform your life by making the right decision after consulting. Therefore, anything that you doubt about tomato production, ask the experts! Are we going to be food secure if we grow tomato?

Food security and what to grow

Food security has been a very big political topic in our country for a very long time now. It has been interpreted in so many different ways. In the next two articles, I will endeavor to answer to some concerns about this topic. I received a lot of questions from an article I had published on this topic in my then Agribusiness Chat column way back in time. A lot of people had asked me what they should grow on their farms to be food secure while making money at the same time. However, some of my readers went on to even accuse me of confusing them because I had earlier written on how lucrative it was to grow soybeans and then later, I wrote about how profitable it was to be producing tomatoes. I refuse to admit that I was confusing anyone including my ardent readers; otherwise what I was trying to do is to unlock their potential by providing as much information as possible so that they can make a good choice as one former finance minister would put it. After reading this short article, I believe that the readers will have the answer of what they should produce or grow. I have reproduced the other two articles in this book for your reference and I hope you will find them beneficial for your farming businesses.

The idea of writing this book was to contribute to agricultural information dissemination. To be specific, I wanted to contribute to enriching the small-scale farmers with the most sought after agricultural information although I know that very few small-scale farmers do have access to books but I believe that in a way or another, this book will be made affordable to everyone. However, I know that there are a lot of people that are working and doing farming at the same time, but lack the basic information about the crops they are growing. It was gratifying to learn that a lot of you pointed out that if you go flat out in growing of commercial crops, then your households will be food insecure. My answer to this is that you were partly right but that is not true. I personally feel that it is more profitable to grow a crop which will give you more income especially if you have comparative advantages in that crop. Why do I think so?

First let's try to define what food security is. There are a lot of definitions and I picked one which defines food security as 'meaning that all people at all times have physical and economic access to adequate amounts of nutritious, safe, and culturally appropriate foods, which are produced in an environmentally sustainable and socially just manner, and that people are able to make informed decisions about their food choices'. So, in food security, there is emphasis on; food availability, access to food, food use and how food is produced. Let me give you an example, some friends from Luapula have been boasting that they are food secure just because they are able to have nshima with fish every day. If you look at the national statistics; you will be surprised to learn that Luapula province is amongst the provinces with the highest malnourished children. One would ask as to why, because it is rare to hear of a time when the government took relief food to Luapula. In short, they have food throughout the year but how come then that they have one of the highest statistics of malnourishment? The answer lies in the type of food they have access to. In Luapula, most of the people eat cassava meal. The population on Islands like *kwa Nsamba* and *Bwalyamponda* only relish they know is fish and cassava leaves. It is rare that they have access to other relishes because of their location unless they paddle to the mainland in Samfya to get supplies. Even if they have enough to eat throughout the year, they are not food secure because they don't have access to adequate nutritious foods. This does not mean fish is not nutritious or cassava leaves, but the two types of foods do not have all the nutritive requirements needed for the balance human growth. Remember at the core of food security is having access to healthy food and optimal nutrition for all. Food access is closely linked to food supply, so food security is dependent on a healthy and a sustainable food system.

Remember we talked of fish being part of the monotonous meal for the people in Luapula province, though this is slowly changing because the waters have been over fished. How is that fish produced? We have heard of the types of fish gears that are used in harvesting the fish especially in Bangweulu. Some fishermen are reported to be using mosquito nets to harvest the fish, and this type of harvesting includes the eggs. The nets have mesh that is less than two inches, this is an unsustainable way of catching fish as it does not only catch small fish, but it also removes or harvests the eggs. In Lake Bangweulu, the fish stock levels have plummeted to the lowest and someone was complaining that crocodiles have turned to killing human beings because there isn't enough fish in the lake. In 2009 there was a project called PLARD which had plans of restocking and I hope they have not shelved the idea. Restocking should come with intensive monitoring if it has to be successful. Is this a sustainable way of catching fish? The affirmative answer is NO, then it means that the people of *Bwalyamponda, Nsamba* Islands and others in Luapula are not food secure. So, then what should the people in such circumstances produce in order to be food secure?

Mind you, we are promoting farming as a business. The first thing farmers need to do is to determine the market for the product/commodity that we want to produce. Take for instance, if the people of Luapula decides to grow cassava because they know that they will sell the processed tubers or mealie meal to people of *Bwalyamponda*, and that they can grow carrots to enhance vitamin A levels for their families. They can also grow rice in the swamps which they can sell at Samfya market or the lodges around Samfya and other surrounding places. How do they decide on the crop/commodity to produce? Once they have determined the market where to sell to, the next thing to look at is the comparative advantage of producing that commodity in that particular area. Comparative advantage (Wikipedia) refers to the ability of a person or a country to produce a particular good or service at a lower marginal and opportunity cost over another. A comparative advantage in producing or selling a good is possessed by an individual or country if they experience the lowest opportunity cost in producing that particular good. Therefore, if cassava is chosen, what kinds of factors of production will they need? No doubt they will need to start listing them and determine the ease of producing it, for instance, land. Do they have adequate land for cassava production and do the soils support the cassava crop? If they note that probably, they don't have adequate land, it would be advisable to choose other enterprises that can give them the highest returns on investments per unit area.

Once they have analyzed all the commodities basing on the comparative advantage, then they will need to conduct an enterprise budget for each of the possible enterprises. In simple terms, an enterprise budget is a listing of all estimated income and expenses associated with a specific production of an enterprise. It provides an estimate of commodities' profitability. This should be carried out on all enterprises/commodities that one plan to venture into because it can help guide them determine whether the commodity they want to venture into is viable. They should also need to have an estimation of the price of that particular commodity and this is one of the difficult parts of enterprise budgeting. The enterprise budget has three parts; the income, variable costs and the fixed costs. Let's give an example of an enterprise budget for maize assuming the price is K65 per 50kg bag so that we can be at the same level in what we are discussing. (It is always good to use the lowest price or price projection for the commodity for planning purposes):[6]

Income:
    50 bags at K65 per bag   K3, 250

Variable costs:
    20kg of certified seed   K 160
    Fertilizers
        4 D Compound at K205/bag   K 820
        4 Urea at K195/bag   K 780
    Labour
        Land preparation   K 150
        Weeding   K 200
        Harvesting   K 150
    Empty bags at K1.50   K 75
    Transportation at 5n/bag   K 25
Total Variable costs   K2, 360

Fixed cost:
    Land on long lease (K2, 000 for 5 years)   K 400
    Total Fixed costs for the year   K 400

---

[6] This enterprise budget was based on commodity prices of 2012

From the above simple enterprise budget, we have seen that if maize was to cost at K65 per bag and we assume that our yields will be 2.5Mt/ha or 50 bags, our income from this enterprise will be K3, 250. We have total variable costs of K2, 360. These are called variable costs because they vary with the level of production and management. For instance, someone will decide to use three bags of urea and three bags of basal fertilizer if the soils are very fertile. However, the fixed costs do not change, whether we use four bags of fertilizer or two, we shall still need to put it in one hectare. You will note that the cost of land is K2, 000 for five years, so to get the cost for each year; we have divided five into the total land costs (simple depreciation). So, the K400 is the cost allocated to land for each year. After adding all the costs, we have come to the total of K2, 760 for the hectare. Therefore, at this level of production, we shall get an indicative profit of K490. I have said an indicative profit because we haven't deducted other indirect costs like our labour and other marketing costs like levies and so on. So, we need to make such a budget for all the crops/commodities that we plan to produce and compare the indicative profit margins of all the enterprises. We shall pick or rank those with the highest positive indicative profits.

Once all the commodities have been ranked that are intended to be produced, we can also carry out an analysis of their nutritive values especially for the small-scale farmers. I am emphasizing on small scale farmers because these get almost all the food requirements from what they grow. If anything, they do less buying. By nutritive value, I mean which foods will give them energy (the energy giving foods like maize, cassava, rice), the protective foods (these are foods which will boost the vitamins and mineral levels in our bodies). These help to protect us from diseases and in this category, are found mostly the fruits. We also look at those foods which will help them grow or build their bodies (high in proteins) such as soybeans, meat, fish and others. We might not produce all these types of crops/commodities on our farms but we should ensure that our households eat all the three types of foods every day. For those commodities that we do not have a comparative advantage in terms of production, we need not produce them but we should produce in large quantities or commercially those that have a comparative advantage and have a positive indicative profit. This is because these are the crops that we can sell for cash and the extra money realized from their sale can be used to buy those that cannot be easily produced. If we follow this type of agribusiness planning and feeding habits, then we are treating farming as a business and will have food security in our households. Therefore, the answer to the kind of crops that one should produce lies literally with oneself. What I would encourage the farmers to practice is growing several types of crops including rearing of different types of livestock to spread the risk. I have learned this through experience and we all knows what happens if the rain pattern is not good.

My grandparents being resource poor small scale farmers were growing several crops (maize, sorghum, groundnuts, pumpkins and watermelons) and were keeping cattle, ducks and chickens. In 1984, the herd of cattle in the kraal had multiplied to over hundred and forty animals. They considered themselves that they had insured us through that investment. They were sure that their grand children will go to school and complete their education through selling of the assets they had accumulated. During that time, animal diseases were rarely heard of. However, in 1985 there was a serious outbreak of cattle disease which was locally called by a Tonga name as *Denkede* (Foot and Mouth) because they suspected that the disease was brought by our brothers who had migrated from Southern province in search of more land for farming. By the fall of 1987, all the animals were wiped out in the kraal. Now imagine if my grand parents had not thought of diversifying into keeping goats as well, they could have been left with nothing. Apparently, this disease never affected goats. This is a very good lesson even today that we should not only grow maize even if the government is providing inputs such as subsidised fertilizers and seed for maize production. Farmers should always diversify or they can divert some into growing of tomatoes, soybeans, watermelons, cabbages and other crops (but not selling to buy *tujilijili* or clothes like some farmers do). It is gratifying that the government is anticipating providing support in other crops like rice, soybeans and cotton. I have no doubt that this will improve crop diversification and ultimately household food security. The answer to the question about what you should grow/produce in order for your household to be food secure remains with yourself because you are the one who knows how much land you have, the resources at your exposure, the market prospects and what your land is capable of supporting in terms of production. What is important is to have a diversified crop production as we will discuss in the next article.

Diversifying for food security

Literally on a daily basis, we hear people emphasize the importance of food security to the economic development of the country or household. Worldwide, food security is a political issue and some people have even lost political power or elections because of food insecurity; the best example being the UNIP government that lost power in 1991 because of food shortages that rocked the country. Leaders that have survived or remained in power when their countries don't have food have done so undemocratically in most cases. There are several cases of such countries even in Europe but that topic is beyond the scope of this book.

In our beautiful country called Zambia, food security is synonymous with the availability of maize being that this commodity is the staple food. Literally, most of the people in the country depend on it for survival. This crop, like earlier stated was introduced by the Portuguese and before that; we used to depend on cassava, sorghum and millet. Since the time, I was a child, my grandparents used to grow sorghum but completely switched to maize production later in the 80s when Zambia had recorded the biggest 'bumper' harvest ever. I should emphasize here that that the three million metric tons (3,000, 000 Mt) of maize bumper harvest that the MMD named as unprecedented in 2013 was not the first as they claimed. We have been having bumper harvests even before but surprising at the backdrop of those harvests have been food shortages and food insecurity at the same time. If you are a logical thinker, you will fail to comprehend that a country that produces bumper crops can be starving at the same time as we have demonstrated with the stuntedness of some children in certain provinces. It is unbelievable! Well, psychologists claim that no one is hundred percent sane but it's the level of insanity that matters. Let us not be philosophical especially that I am not a psychologist although I should be haste to state that the environment is slowly making many of us to become just that. Citizens have been subjected to conditions that is driving some to insanity.

When one produces maize, he/she can only claim the volumes that he/she is able to convert into mealie-meal or cash. What is the meaning of this statement? The Lamba's have a saying which goes like *'akafumbe kantu ufumbete-muminwe'* (literally meaning, the mice you own is the one in your hand). You can't find two holes for the mice in the ground and claim that you already have them or count that you have two mice. We have been producing good harvests for the past four seasons since 2010 but even at the backdrops of those bumper harvests, we have had people starving and many to the extent of being malnourished. We grow enough maize to go around all people in this country but our grain management has been pathetic to say the least. In the first place, we do not have adequate storage facility as a country to securely keep all the maize we grow.

At one time, we had neglected the multi-million-dollar storage facility like the one at Kabwe's Natuseko depot to dilapidate to levels where they were breeding grounds for snakes and bats. Meanwhile, our maize was being stored under plastic tarpaulins which easily got damaged with wind and rains. The commodity that can make a politician win an election was stacked on logs and once the logs rot, the crop was soaked leading to it being disposed-off in an unenvironmentally unfriendly manner like what happened in (2011/12) season in Petauke and many parts of the country. In the fifth national development plans (FNDP), it was planned that a lot of the storage facilities would be rehabilitated and a lot of them where going to be built under FRA. Trillions of kwacha were budgeted but none of it was released and if any was released, it was misapplied. In the last eight or so seasons, we have received relatively better rainfall without any serious challenges of diseases and pests until the year (2012/13) when we have been hit with the devastating army worms. What implications has this got on our food security?

Firstly, from 2013 to 2015 we had late onset of the rains for the main season. The rains had come slightly late meaning most of our medium scale farmers that had been growing late maturing varieties which normally have yield potential of over 10Mt/ha had opted to plant either medium or mostly early maturing varieties. These varieties have yield potentials ranging from 5 to 8Mt/ha. No small-scale farmer grows a variety and attains the maximum yields as prescribed by the potential because of various reasons ranging from poor weed management, lack of fertilizers as they are supplied late by the government. For instance, in some cases they start by supplying urea in December and basal fertilizer in January. Poor rainfall distribution, lack of good soil management practices, and low plant population per unit area amongst some other factors which contributes to low productivity or not yielding the potential of the seed varieties. With the outbreak of the Fall army worms in 2012, farmers observed increased prices for maize, later on mealie meal. Politicians dispelled the assertion with the contempt it doesn't deserve because they did not want to bear the brunt but that was reality. Some agricultural experts were claiming that the Fall army worms have been controlled when they had completely wiped out the crop in some cases and had transformed into pupa in readiness for another break out in five to ten years later. One farmer in chief Chipepo's area of Kapiri whose crop was eaten by the army worms was asking where they could find extension officers in their area. The next nearest extension worker resides at Luanshimba 60km east of where this poor farmer lived. Meanwhile, in Lusaka the vice president then, Dr. Guy Scott was launching the replanting season with the seed donated by Pannar seed. By the way, much of that seed was given to farmers along the line of rail while farmers such as those in far flung areas like the shores of Lukanga swamps, the 'belly' of Zambezi river in Mbanga of Lukulu districts and Vubwi had a raw deal. With such new pests like Fall army worms attacking our crops, we need to come up with new production dynamics in agriculture.

The changing production dynamics

In not more than thirty five years ago, the former minister for Southern province in the late president Sata's government used to produce more maize at his farm than the combined production of Luapula and North western provinces together. During that time, Southern province was the grain basket of Zambia. The production dynamics started to shift around the late 80s to Central province and in the last four seasons, Eastern province has been on the lead in maize, groundnuts and sunflower production These are the crops which are predominantly grown by the smallholder farmers in Zambia. For Eastern province, production areas range from Petauke, Katete, and Chipata going up to Lundazi. However, in the last two seasons we have noticed the northern circuit increasing in maize production especially Muchinga and Northern provinces. Southern province has been drier than ever and a lot of our brothers from that region have migrated to Central and Copperbelt provinces in search of land which receives more rain and is fertile.

This is critical information which should be analyzed well by the ministers of agriculture and the counterpart at national planning. The reasons being that certain farming systems and crops are at risk of reduced production. I have in mind cassava and millet which are crops predominantly grown up north. Cassava might not be a very important crop as far as the market is concerned but people that have lived in Luapula and Northern provinces would agree that it is a very important crop for them. As maize production substituted cassava and millet production in that part of Zambia, we are likely to see increased instances of the people asking for relief food from the government. For the three years, I had lived with the people of Luapula, there was not even a single year I heard people of that province asking for relief food. This is because of their production systems and the types of crops grown. They could have high levels of unbalanced malnutrition but they are always household carbohydrate secure. We know that cassava though a heavy feeder, it can be grown even in less fertile soils and it is an efficient user of soil nutrients. The team leader of the Root and Tuber from ZARI with his team members have done remarkable research on breeding high yielding cassava varieties which matures early as opposed to the traditional varieties.

As the cereal production belt shifts up north, the government and its partners should start thinking of introducing cassava and sorghum production to drier areas that are being abandoned. This will enhance the food security of the people in those areas. The other factor that needs to be considered is the creation of the market for the crops grown. Luapula has sustained to be growing the crop because of the market from D.R. Congo; there are a lot of Congolese that cross the Luapula River and Lake Mweru to buy the commodity from the farmers in Zambia. I know that Angola is also another huge market for cassava and with good infrastructure such as roads, production of cassava in Southern province can have a ready market and this will make the farmers grow it more. The other opportunity we need to explore is to compel industries such as Zambian Breweries to be using cassava in its brewing of the alcoholic beverages[7]. In 2009, I started negotiations with Zambian Breweries through their representative in South Africa for them to be buying cassava. The challenge we had then was to quantify how much cassava was produced in Luapula but the negotiations had continued. However, in 2011 I got information to that effect that the company was putting up a plant on the copper belt which would be using cassava as a raw material. The other opportunity we need to explore is the use of the same commodity in bioethanol production. I am alive to the fact that in Mozambique they are using the commodity as a raw material for this purpose.

Indeed, as the southern parts of the country becomes drier the grain production is shifting up north which is a great threat to the crops mentioned. The country should not at this point dare to loose cassava production the way sorghum production has been abandoned because of heavy promotion of maize cultivation. Remember that as early as 1985 people asked the government then to be investing in energy but because we had enough energy then our leaders thought it

---

7 Zambia Breweries has since started using cassava in its brewing of beer

was not necessary and went to sleep without investing in more power generation. I wouldn't want to have a similar situation twenty years from now such that we will even start importing cassava cuttings to reintroduce the production of the crop. The Bembas say, *'ukwali insoke takwafwile muntu'* (when you are warned of the danger ahead, you will be aware and take precautions). However, we know that to grow cassava for a family of six, you need adequate land. For maize, half a hectare will be enough to grow enough to feed such a family and sell a few bags. Cassava will need a bit more land than that, meaning land is an important factor in agribusiness. However, the current land available in Zambia is more than adequate to feed the entire region if we put it to good use and stop the deforestation and bad agricultural practices that are currently taking place.

Importance of land in agriculture

In the earlier article about agribusiness development, I picked out two important points: the importance of having land and the value of having the right marketing information. I'm compelled to write about this important factor of agribusiness development because of the way we have been misusing land in this country. My grandmother once told me that you can only know the importance of something when you lose it. This is during the time we were misusing water drawn from the well in the village. I used to argue that water is readily available and we can afford to even pour it anyhow, but she would rebuke me that there are some people in the world that are buying water and worse off, fighting for it. I did not believe this because I was young, not until when I grew up and started to appreciate her guidance. This is true with land, we seem not to care much about land in this country because we seem to have a bit more of it. Just across Zambezi river, if one was to ask our brothers in Zimbabwe, the land has made them go to war amongst themselves. Even in our country, we have watched on television and read in newspapers people that have been killed over land especially here in

Lusaka and Kitwe. If anything, most of the wars that have been fought has been caused because of land.

Not long ago, we thought we had so much land in Zambia such that we did not care whether people practiced *chitemene* system of cultivation or *Galaudza*. People could cultivate a piece of land for only three years and abandon it for another virgin piece of land. Sad enough, in some parts of Zambia, they are still following such detriment practices. However, it has come to the fore that amongst the factors of production, land is the most limiting. One can increase his/her human capital by acquiring new skills through trainings; one can increase the financial capital by acquiring a loan but this is not the case with land – if you have one lima (0.25ha) of land, it will remain just that. It is very difficult for us especially that we are landlocked to expand our land. Our friends that have access to oceans might dream of reclaiming land like the Dutch have done, though it is a very expensive venture. May l ask those that are still practicing *chitemene* system to visit their brothers in Eastern province and learn how important land is, especially those living in Chipata, Katete and Nyimba. Some barely have a hectare which are even stony. Not only from the Eastern province have had we got land pressure but other provinces as well such as Southern and Central provinces. A lot of my brothers from those areas have even migrated to other regions such as Northern and Copper belt provinces in search of this scarce commodity. I would not be surprised to learn that some have even ended up crossing into Congo D.R.

I am encouraging those that are still using the old ways of making ends meet from land to adopt modern farming practices such as the conservation or smart agriculture. A lot of farmers in Eastern, Central and Southern provinces have adopted these farming technologies and have been sustainable in their production. There have been a lot of success stories with conservation agriculture that can be shared in this country. A lot of people though, have a lot of misconception about conservation agriculture by thinking that it is the digging of basins alone. A lot of lazy people like my cousin think he cannot manage to dig the basins – NO! There is more to conservation agriculture than digging the basins (potholing, utulyompi, mahenje, utumpenshipenshi). There is a lot of equipment that have been developed depending on the standard of agriculture that one is practicing. I should commend CFU for the commendable job they have done in this area of land management. It is very disturbing that even some quarters of the ministry of agriculture had some reservations initially about these technologies, but they have worked wonders for those that have dared to adopt them. We have implements like the famous Chaka hoe for those that do not have animals drawn implements. Studies have been done and proved that if one works for an average of two hours per day to dig the famous basins and starts his work immediately after harvesting (May) when the soils are still soft, he could cultivate up to five hectares by October. To make the work easier for those that have oxen, a ripper/planter has been developed which can have a hectare ready in an hour's time. For the commercial small scale farmers that have tractors, there is a very good learning school of conservation agriculture at intensive commercial level – York farms in Lusaka. That small but productive farm has been under cultivation for over thirty years and the land is always carrying a crop every time of the year.

What we should all bear in mind is that with the current global warming and climate change, some areas will be too dry for our crops while others will be too wet; this applies to seasons too. Therefore, we need to adopt technologies that mitigate the effects of climate change and its impact on agriculture. You will agree with me that whether it rains or not, we will continue to eat and how we do that calls upon every one of us to respect land. This also goes to our traditional leadership that are in a habit of selling bigger chunks of land to foreigners for a song to rethink and stop at once. The government too, should start to enact laws that limit foreigners to own not more than five hundred hectares (500ha) of land and no individual should own more than 5,000ha. We should seriously start to think of redistributing land to indigenous Zambians otherwise our children will be found in the predicament in which the Zimbabweans are. Land is a very important resource and factor in agribusiness development.

The second point to note is the value of market information. Amongst the factors that make our rural smallholder producers' commodities to be uncompetitive is lack of market information. Timely marketing information provides the basis for decisions making such as pricing, distribution and promotion amongst others. When I wrote an article that was discussing about producing crops that may give you a bumper 'income,' the questions a lot of my readers asked is where they can find information about what crops or commodities that will be fetching better prices that marketing season. Many of you will agree with me that farming has become as sophisticated just like mining. Actually it is more than mining because in agriculture, you are dealing with living things. It is very difficult to manage things that are living than non-living things. If you know the better combination of cement, sand and water when making blocks for construction in Mufulira, you can make the same blocks to construct a house in Vubwi as long as you know that the environmental conditions are similar. However, that is not true with agriculture. It is the reason that teachers, mentors and coaches of agriculture are emphasizing practicing agriculture as a business. Don't be cheated that by adding black soil to your lawn, then it will grow healthier. The black soil has to contain the right amount of plant nutrients to support the healthy growth of that lawn.

Well, not delving deeper in the live science of farming, allow me to discuss the social science of farming. I believe we all agree that farming is no longer a livelihood but a business. If there is anyone of us with an iota of doubt, I will encourage them to take a trip and go to Namwala district. You will be amazed at the number of people that are driving fancy cars, better than the ones we drive; the so-called managers in towns. Additionally, compare the economy of Chipata with that of Mansa, you will agree with me that Chipata is far much better and developed than Mansa even though both towns' economies depend on agriculture. What is it that the other town has done well than Mansa? The simple answer to this research question is that Chipata has used the power of information well by adopting the concept of farming is a business. The difference between the two communities is that majority of the people in Mansa may have data about farming as a business whilst their counterparts have got information about farming as a business. You may want to differ by asking a question as to what the difference is between data and information. In a layman's language; information is processed data. Let's discuss using the following example.

Most millers and feed manufacturing industries in Lusaka have been looking for soybeans and they have been paying relatively better prices for those commodities. Kabwe is about 135 km from Lusaka while Chipata is 580km. Both provinces grow soybeans but from the analysis that I did when I visited one farmer friend who managed to harvest 2.5Mt (50 x 50kg) of soybeans, he could have made more money by him selling directly to the millers. This friend sold the soybeans at K2.10/kg to some traders that took the soybeans and sold it at K3.00/kg in Lusaka. This friend did not analyse how much it was going to cost him to deliver his soybean to Lusaka. The trader paid K0.30/kg for transportation and handling fees, meaning this trader made a profit margin of K0.60/kg. Now suppose this friend decided to take the soybeans himself to Lusaka, he could have made an addition K0.60/kg to what he got. I would not have an issue with him if he was involved in other enterprises such as horticulture production; I would have excused him to assume that he could have been busy with other crops but this was his sole cash crop that he grew. To date (October), he hasn't been doing anything apart from waiting for the rainy season to come so that he can plant the next crop again. It is important to shorten the value chains from the producer to the processor in commodity marketing. However, many times this is not possible in Africa because of the imperfect market systems.

On the other hand, I had visited a different farmer in Chipata district. Immediately, I arrived at his homestead, I could tell that the farmer was in business. When I compared the amount of land this farmer was cultivating, it was a third of what the farmer from Kabwe was cultivating. This farmer was involved in growing of half hectare of maize for consumption but besides he was also growing cotton, groundnuts and top of all; onions. It was interesting that this farmer prepares his nursery to be transplanted in November. He has employed workers that help him to look after the nursery which is normally prepared in May. He hardens his seedlings and removes them from the ground in September and stores them in a shade before transplanting in

November. Once that is done, the farmer starts to harvests his onions in February when the country has less supply and resorts to importing onions from South Africa to meet the demand. This farmer does not need to transport the onions to Lusaka because customers from Lusaka and Kasumbalesa travel to buy from him at the farm once the crop is mature. Many traders from Soweto and Kasumbalesa which are huge markets for vegetables knows him for this unique type of farming system he has innovated.

From the two examples of rural farmers I have highlighted, you will agree with me that the Chipata farmer has processed the data about the market and transformed it into market information which he has used to his benefit. Unlike my Kabwe farmer friend who quite well got the data and never did anything to make it a meaningful market information. I decided to give these two examples to try and elaborate the importance of market information. We can enhance the flow of market information by improving the communication infrastructure. I forgot to have mentioned that the mobile connectivity for the Kabwe farmer is quite poor because he has to climb a tree to access the network while the other farmer can access it even in his bedroom. However, I should emphasize the point that no one is going to get market information for you as a farmer; you need to search for it because you are in the business of farming. It is therefore important to emphasize that farming is a business and market information is very valuable tool in any business. Therefore, is it possible for one to make money through soybean growing in Zambia?

Soybean is 'a money spinner'

We all know that agriculture is not only one of the biggest sectors in the country but it also contributes immensely to the gross domestic product (GDP) for Zambia. It contributes over twenty percent to the country's GDP, and over eighty percent of the actors (farmers) are smallholder in nature. It is therefore prudent that we discuss the business of growing soybeans in Zambia.

Soybean is a crop that belongs to the same family as beans, tomato, eggplant and many other crops. They are collectively called leguminous crops. Unlike maize, the bulk of soybeans consumed in Zambia is grown by the commercial and some of the emergent or semi commercial farmers who I have chosen to call as commercial small scale farmers in this book. It is a crop that is not too difficult to grow if one judiciously follows its requirements. It is not by design that smallholder farmers don't grow much of it, but some reasons have to do with the way it is marketed. However, with problems of late payments in maize marketing by Food Reserve Agency (FRA), there has been an upswing in the adoption of soybeans cultivation by smallholder farmers because payments by the processors is on the spot as the commodity is delivered. For instance, in the 2011/12 planting season, no seed suppliers had certified seed by November 2011 and this was the case in 2015/16 farming season. All the stocks had been bought due to the high demand for this commodity, unfortunately not only in Zambia but the whole of Southern Africa region.

Soybean belongs to the family *Leguminosae* and genus *Glycine L*. It is a crop native/indigenous to East Asia (China), which is one of the five oldest crops cultivated by the Chinese for food before 2500BC. It is a crop that can be grown in a wide range of soils, though it does very well in deep well drained sandy loam to clay loam. For economic yields, avoid growing it in sandy soils. However, the soils should also not have a pH less than 5.5 (pH is scale on which we measure soil condition in terms of acidity. The scale is from 0 to 14, with values near zero being very acidic while the opposite is true for 14. Soils that have pH 7 are neither acidic nor basic, they are said to be neutral, 0 to 6.9 is very acidic to slightly acidic while 7.0 is neutral and 7.1 to 14 is slightly basic to very basic). Preferably, soils are recommended to have a pH range of 5.5 to around 7.0 because that is the range which promotes most crop production. Though I should mention that it is relatively cheaper to reclaim a soil that is acidic unlike one which has higher pH values, and lucky enough most soils in Zambia have soils that are acidic especially in region III. The topic of soil acidity will be dealt with later in this book.

Therefore, the range of 5.5 to around 7.0 would be the ideal range for our soybean production. The problems with soils that are very acidic are that some elements that are important for soybean growth will not be in the unavailable form for the crop to access them. One of the elements which are critical for soybean growth is phosphorus (P) and at very low pH, it is found in amounts that are not available to the crop (complexed form or fixed in soils). I will not delve into the science behind the availability as this could be too detailed for this space. Some people might be tempted to say even if the soil is quite acidic (lower than 5); I will apply to much fertilizer. My advice to such lines of thinking is that you are likely to just waste your resources because it is like putting one table spoon of sugar in five liters of water and expecting the solution to be as sweet as *zigolo* (sugar solution) or worse off putting sugar that is wrapped in a plastic in water and expecting it to dissolve. However, for some farmers that might be found in soils that are in such conditions, the remedy is to apply lime (mind you, one can find very acidic soils even in region I), it doesn't just come about due to high rainfall but also the agricultural practices such as continual application of acidifying fertilizers like urea and ammonium nitrate for a very long time without liming.

This crop also needs rainfall ranges of between 500mm to around 1000mm depending on the variety that one wants to plant. The rainfall has to be well distributed. Our farmers are encouraged to seek advice from agricultural extension staff from the ministry of agriculture, ZNFU and some private companies such as SeedCo, MRI, ZamSeed, Pannar, Pioneer, BASF and many more. The crop also needs a frost-free season and do not need to be grown in areas where temperatures go beyond $40^{0}C$ for a long time. With this information, you will see that it can be grown in a wide geographical coverage of Zambia. In areas where rainfall is not well distributed and falls on the lower bracket, irrigation can supplement its growth if one can afford.

After being satisfied by the environmental conditions, the next thing that the farmer who we shall be calling a 'green entrepreneur', needs to do is search for a suitable variety in the area he/she is found. Just like maize, we have varieties that are early maturing and those that are late maturing. The other very cardinal point to note is that some varieties do shatter (explode when they are mature) while others take long or may not at all. This is a very important characteristic to know because if one plants a shattering variety and goes in late to harvest, they will have high losses. Planting soybean can be done on ridges or flat land. You can also use conservation methods in planting this crop and the plant specifications depends on the level of management one will implore, however, farmers should not plant less that 80kg per hectare if one wants to get near the yield potentials of most varieties. The recommended rates for most of the varieties are from 80 to 120kg of seed per hectare, and the newly released varieties have yield potential of between 3.0 to about 6.0Mt/ha as compared to old varieties that would yield around 1.5Mt.

The PF government came into power on the promise of more money in people's pockets. So, if you want more money in your pockets, you need to plant improved varieties that will give you maximum yields. At planting, there is some material that you can plant with soybean seed; this is called an inoculant. Inoculants are not fertilizer so to speak but these are small living organisms (bacteria – harmless to the plant) which when well applied will help soybean use the atmospheric nitrogen and convert it into plant usable nitrates. We all know that nitrogen is one of the most important macro nutrients needed for not only soybean growth but even other crops but yet so limiting. A word of caution is that this inoculant should not be used to substitute the fertilizer because then we shall starve the plant off other necessary nutrients like phosphorus, potassium, boron and many others.

All those that want to venture into soybeans cultivation should take note that unlike some crops, it does not compete well with weeds in the early stages of its growth (first four weeks). However, we are lucky that just like maize, there are herbicides that one can spray in order to control the level of weed infestation. There are some herbicides that can be sprayed before germination and others that can be sprayed after germination (over the soybeans). Farmers need to visit various chemical companies to be advised on what kind and type of herbicides to use on particular soils. What farmers need to know is that there are about three weeds species that are quite notorious in soybean production. Some herbicides on the market have failed to control them and these are *Commelina*, *Euphorbia* and wild cucumber or wandering jew/morning glory.

The types of weeds that are predominant will tell what kind of herbicide to apply, is it broadleaves or grasses? BASF, a chemical company has some great products such as Frontier, Optil and Hammer that can be sprayed pre-emergence to controls grasses and some broad leaf weeds while the latter controls mostly the broad-leaf weeds. They also have Basagran which is sprayed post-emergence and controls mostly broadleaf weeds including the notorious yellow nutsedge. Syngenta have Fusilade, Bateleur Gold and Flex as well as Dual Magnum in their arsenal for weed control. Other herbicides that can be used are Classic and Strongarm. In 2005 while visiting a farmer who was complaining that the herbicide he had applied was not effective as the weeds where not controlled, I leart that the farmer had not followed the instructions on the label. After a thorough inspection of his field and investigating, it was learnt that the farmer had applied the right type of herbicide but in a wrong type of soil. The clay content will determine what type of herbicide to apply and at what rates - this was not followed. Some herbicides do not need to be applied together with organophosphates or carbamates for instance, and most farmers don't follow this. Just by forgetting to thorough wash the sprayers after using them with insecticides like monocrotophos or chlorpyrifos which are organophosphates can change the efficacy of the herbicide.

Just like people, soybeans are living things and they can get diseases and die. There are so many diseases that affects soybeans but the disease infestation is not so much pronounced with the smallholders as it is with the commercial farmers. The one major reason for the disparity is that commercial farmers follow intensive type of cultivation due to limited land unlike small holder farmers who may leave some land that they leave to furrow for some time. However, smallholder farmers must also practice better land management as taught by our colleagues from CFU; Zambia's land is limited. Soybean is affected by both fungal and bacterial diseases such as rust, powdery mildew, and bacterial blight, red leaf blotch and others. There are also several pests that affect soybeans, such as caterpillars, aphids, cutworms and other sucking pests as well as birds. There are some chemicals that may be used to control these except for birds (please avoid bating birds as you may end up killing other people, just scare them away). The best preventive measure to control pests and diseases is to follow what I call field hygiene – which starts with proper crop rotation, physical barriers and use of recommended environmentally friendly chemicals.

There was a time I attended a field day organized by some agriculture firm. One farmer asked the coordinator the importance of growing soybeans. I could see the coordinator bubble with confidence when answering to the farmer. His answer was like 'the importance of growing soybean is that it puts money in farmer's pockets, one can make cooking oil, and you can make *nyamasoya* (soya chunks) and high energy protein supplements (HEPS) for those that have HIV/AIDS and nutritional purposes'. After the answer, everyone went like woo! He almost went on to the next crop and at that point I thought the coordinator was not doing justice to one of our stakeholders in crop productivity. With all the respect, I interjected that in addition to the many things that the officer has mentioned; the other importance is that soybeans help to fertilizer the soils. Remember that we mentioned that soybean is one of the few plants which fix atmospheric nitrogen into nitrates (plant absorbable form of nitrogen). This is very important in that it helps to enrich the soils with nitrogen which is one of the most important macro nutrients for plant growth. If a plant is lacking nitrogen, it will turn yellow and this will reduce on the yields, quality and in severe cases, there will be total crop failure. When we apply urea or ammonium nitrate, what we are actually applying is nitrogen, so soybeans are a very important stakeholder in improving soil productivity as it helps to fertilize the soil with nitrogen (the process of nitrogen fixation is beyond the scope of this book).

I should state here that before even thinking of growing soybean, what one needs to find out about the crop is the market. However, I took it that farmers of nowadays know what is supposed to come first. Nonetheless, we are repeating probably one of the messages that have been repeated several times by my colleagues from the agribusiness department of ministry of agriculture and livestock that farming is a business, and a business involves buying and selling. So, the important thing to know about soybean is who is going to buy my crop and at what prices will they buy it? Unfortunately, this crop is not part of those that government supports through FRA and I have failed to understand why, but I don't insinuate that they start buying as they might just distort the market forces. I would love them to support its production the same way they do with maize. However, we shall deal with that later as we build our case in this book. I have met some farmers who when you talk to them about a market, the only thing they think about is selling to Europe or abroad (they are like my friend Katungu Mukelabai who only thinks that a market is only when one is able to sell to Queen Elizabeth or South Africa). However, in the 2015/16 the government has provided for e-voucher in addition to the convention FISP and this will allow farmers to buy inputs of their choice such as soybean inputs? This is a great innovation from the government and we salute them for the effort.

Techno Serve Inc. is an NGO that was promoting soybean production and marketing in Zambia and Mozambique. According to the value chain studies that they conducted, soybeans have a ready market in the region and the largest quantities go to South Africa. For Zambia, we have been utilizing all the soybeans that have been locally produced and have been exporting little surpluses that we used to have except for 2011 when we needed more than we produced. This doesn't mean that there has been a year when we had produced more than enough and failed to sell it like has been the case with maize. The entire soybean that is produced is bought off from the farmer. What my friend Katungu and some other farmers need to know about marketing of soybeans is how much are they supposed to sell the soybean at. However, in 2012/13 we seem to have produced more soybeans than we needed in that there were still some soybeans with farmers even the time I was writing this chapter in September of the marketing season. So where can we sell our soybean to?

In Zambia, we have a bad term we call the SME buyers of agricultural commodities as briefcase buyers, scrupulous dealers/buyers, swindlers and so on. I see these people as very important businessmen and women because they provide a service that is very important to complete the value chain – the market. Just imagine, if we didn't have such people, who was going to be buying all the soybeans, cassava, groundnuts, tomatoes, sweet potatoes, name them, that are produced by farmers since FRA doesn't buy all these commodities? What I think we needed to do is to educate, train and mentor our smallholder farmer on how to do the costing and price risk mitigation. Though I feel the biggest challenge is on extension side of the value chain. In February of 2013, I wanted to buy a goat as a present for my small boy who likes pets so much and that he wanted a puppy. As I'm not a fan of pets, I opted to get him a goat so that he can fatten it in readiness for the big feast later in the year when he is tired of herding it. I went to some village, and two young men in their mid-20s brought two goats. When I asked how much they were selling, they threw back the question to me and asked how much I was able to pay for one.

Knowing the prices which most farmers were selling such sizes of goats at that time, I told them K150 (knowingly that the right price was K250) and they told me to at least add a K20. They were very big goats but don't ask me how much I ended up paying because you will think I'm lying though I sat down with them and asked about the costing and how they can go about them. One of the problems that make costing difficult is the attitude by the small-scale farmers of not keeping records. What am trying to drive at is that the market for soybeans is readily available but what farmers need to know is how to do the costing so that they can arrive at an economic price which will leave them with a profit. We have processing companies like Zamanita[8] that have since increased their processing capacities by putting up more processing equipment, we also have organizations like COMACO that is adding value to various agricultural commodities, we have so many grain trading companies and individuals. Cargill Zambia, a company known very well in cotton business has gone into an arrangement with its farmers where they are

---

8 It has since been bought off by Cargill, and at the time of publishing this book Cargill had closed shop

promoting crop rotation of cotton with soybeans, and they are buying off the soybean as well.

The important attribute of soybean is that there are so many products that can be made from the it. In Zambia, the value chain for this commodity is mature in that we have both upstream and downstream industries. It is both vertically and horizontally integrated. For instance, you can think of processors that are making poultry feed who happen to be the major consumer of the crop, we have those that are extracting oil for cooking, we have those that are making health products like soy chunks for those who claim to be vegetarians, HEPS, and baby foods to combat malnutrition, the cake itself for feed formulation, in short nothing is thrown away as a by-product from soy seed. In countries that have developed the bio-fuel industries, they are even making biodiesel from oil that is crushed from soybeans.

The other major problem that smallholder farmers need to work on is to improve productivity besides the costing problem. An average yield for a small-scale farmer is 0.9Mt per hectare (though of late it has increased to 1.2Mt/ha), this is a very low yield as compared to the yield potential that range from 3.0 to 6.0Mt/ha. Amongst many factors leading to this has been low seed rate per hectare, poor land management, use of recycled unclean seed, non-application of fertilizers and inoculants, too deep planting of the seed, poor weed management, and disease and pest management or in general bad agronomic practices. So, when it comes to pricing, the farmer would want to recover all the costs incurred per hectare from 0.9Mt of soybeans, making his/her price more expensive than not only on the local market but the world market too.

From some studies conducted, it has been established that the gross margins for a small-scale farmer are higher than that for a commercial farmer. Therefore, if a small-scale farmer got yields of as high as 2.0Mt/ha, they may make more money per unit area than a commercial farmer who has got yields slightly higher than this due to high variable costs associated with commercial farmers. The average yield for commercial farmers is around 2.5Mt/ha. Some people might be wondering as to how they can manage to harvest a field of soybeans by hand if they want to have more money in their pockets by increasing the area under cultivation. I happen to receive good news for the small-scale farmers that there is some tractor mounted soybean harvesters that can be procured from agro companies like SARO, CAMCO and others. This is a harvester that needs about 75HP, and for those that have tractors and are residing in areas where there are a lot of farmers growing between one to five hectares of soybeans, it's an opportunity to get this equipment and start harvesting the soybeans for farmers at a fee. It is more expensive to hire a combine harvester to harvest a hectare of a soybean field than it will if one used a tractor mounted harvester. Entrepreneurship starts here; besides you will use that harvester for your field as well.

In concluding the topic on soybeans, I would like to let all those that have been asking where they can take their soil samples for determination of various parameters that they can see the agriculture extension officer (CEO) or subject matter specialist near their areas in their respective districts for advice or visit Mount Makulu in Chilanga, Misamfu Research station in Kasama, Msekera Research Station in Chipata or the University of Zambia, School of Agriculture Sciences (Soils Department). It's also important that before farmers apply any chemical, they need to understand very well how it works, read the label and ask the person who is supplying that product to make them understand it and if possible seek a second opinion from an experienced farmer from their area that could have used the product before. Soybean is indeed a money spinner.

To sum it up you can position your soybean production for money, how? A lot of farmers grow crops without understanding why they are growing them. Soybean, even though introduced in the country a couple of decades ago, had not been accepted by the majority until a couple of seasons ago. Now that soybean has been adopted and is being grown by smallholder farmers in Zambia, we will have constant supply of the product on the market. Allow me to share a few tips about successful soybean production. Firstly, we know that soybean, just like groundnuts, is a perfectly suitable crop for crop rotation. Soybean is in a family of legumes, crops that have the capability to fix nitrogen. Soybean can be grown all over Zambia, but it does not do so well in sandy soils. If one has to get a good crop, planting has to be done by mid- November to mid-December depending on the variety. The recommended seed rate for planting ranges from 80kg of seed to 100kg though some farmers go as high as 120kg per hectare. Some commercial farmers have gone up to 130kg per hectare but this tends to have a negative bearing on the yield because of plant etiolation as such rates give a plant population as high as 450,000 plants. Soybean seed loses viability quickly and a germination test as close as possible to sowing time is essential if one is keeping his own seed, otherwise it is recommended to buy fresh seed every year even though it is an open pollinated crop. Earlier in this book, we discussed inoculation but probably what we can emphasize is that for nodulation to be effective, we need not apply a lot of nitrogenous fertilizers. However, the crop has a high requirement for phosphorus and it is the reason why we need to watch the soil pH and probably provide half rate of recommended basal fertilizers. Some companies have formulated soybean specific fertilizers that have a low nitrogen rate up to seven percent. It has an extended period of P uptake right up until mid-pod fill.

On the market, there are some specific fertilizers formulated for soybean with high P ratio. At planting the soil temperature needs to be above 14°C for successful germination but above 25°C is optimum for rapid emergence. Sowing depth is important for good establishment, too shallow will lead to desiccation of the seed but too deep can mean seedlings failing to emerge, and 3cm is usually optimal. Soybean is photo-sensitive, meaning that when the nights start getting longer, it triggers a point for the soybean plant to change from the vegetative to the reproductive phase of plant growth. This is one reason you cannot grow the crop successfully in winter. If grown in winter, you may have small pods and this will impact on your yields. The crop requires heat to develop, grow, and mature. The effect of this heat is cumulative as the growing plant progresses through its life cycle. The crop, just like any other plant, is affected by so many diseases and pests; some of the common diseases are rust, frog eye, leaf blotch, and powdery mildew. It is one of the few crops that you cannot successfully grow on a large scale without the use of fungicides especially in a very wet season like we normally have up north. As your partner in agribusiness, I will not stop warning you of what you may expect in the production of this important crop. As I always say, I will not keep from you some of the magic activities you may need to do in order to get good yields. Commercial farmers are now talking of average yields of between five and eight tons. As a smallholder, there is no reason you can fail to get three tons per hectare. The difference is that in addition to weed control and inoculation, good farmers are using fungicides such as Opera, Amistar Extra and others on the market. Opera is a very good product which goes beyond disease control but works on three building blocks. These are: the product will increase plant growth efficiency resulting in more efficient photosynthesis, excellent wide disease control and increased tolerance to stress. This is what will give you the differential higher yields, improved marketable quality ultimately leading to higher return on your investments. The formulation of the product is so effective that my baby brother in the village last season managed to get a yield difference which has made him to appear as though he is an 'expert' farmer amongst his smallholder peers. The owners of this product have come up with an even better product with excellent formulation and they are calling it as Priaxor. It is worth trying these products because they can make you become an excellent farmer.

In a promising wetter season like the one we had in 2016/17, do not gamble with your investments because it is suicidal. Like I told you some years back in my articles about maize business, soybean will still be a profitable commodity for a number of marketing seasons ahead because the challenges of disease management in a wetter season will separate real farmers from jokers just like the drought of 2014 did. Agripreneurship is no longer a game of chance; it is how you invest in proven technologies. Nonetheless, it is not only soybean production which can bring more money in the farmer's pockets, there are other tested crops like onion.

Onion, the 'Sulphur' mine

In my fifth grade at primary school, I learnt about a balanced diet as food that consists of carbohydrates, body building foods or proteins and vitamins. In Zambia, we have been emphasizing too much on taking so much carbohydrates at the expense of other foods. It could be the reason that we are having a lot of people with pot bellies. It is even dangerous if one is taking too much carbohydrate with less exercising. This trend has even affected the way we are promoting production so much of certain crops at the expense of others. In this article, let me share an opinion on onion production.

Indeed, of late there has been hype for people to grow tomatoes just because one fruit in Lusaka is costing K3 on average and a box at between K250 to K300. As of 2015 many of the seedling producing companies had run out of tomato seedlings. This is not strange because many times in the rainy season, we have seen farmers' gold-rush for tomato growing due to the relatively good price that it fetches. It is not that the population increases in the rainy season hence, the increase in the price but the quantity and quality of tomatoes on the market reduce because of many factors; one of which is that it is very difficult to manage a tomato crop especially when grown on open fields for many farmers. However, there is one crop that has received less publicity but it also fetches good money during this period of the year. Onions; have you ever thought of how much an average bulb of onion costs?

Let me share with you; almost every year between January and May, Zambia imports onions from South Africa and sometimes Tanzania. There are only two to three farms in Lusaka that I know to be growing onion on a commercial scale and their target is the export market although they supply the local market as well. Onion is predominantly a crop grown by 'real' smallholder farmers who either use treadle pumps or watering cans for irrigation. With that background in mind, it is not very common to see even half a hectare of a crop in Zambia, as opposed to tomato where you will find some smallholder farmers cultivating up to six hectares of the crop especially in areas like Mkushi and Lusaka west. Most of the onions are grown between July and November, and immediately the heavens 'open', farmers abandon its production to grow maize crop. This tends to leave a void is as much as supply is concerned. It is true to insinuate that for every tomato fruit consumed, there will be a bulb or half of onion too but then why haven't we publicized so much the production of onions? If one had a hectare of ready onion between January and March, he is likely to buy a canter or a luxurious Jeep from its sale. I remember of a couple of seasons when we had so much tomato in the rainy season such that the box of the commodity was fetching as low as K10 a box (20kg) in real money terms but there has never been such a season when we had so much onions that the price had dropped so drastically.

It's amazing at how we let opportunities go untapped and yet we cry that we need a market for our produce. Indeed, we will continue crying for a market if we think farming is only growing maize and tomatoes. For your own information, whether it is in the rainy season or dry season, onion is always sold per bulb while tomato is sometimes sold as a heap. The K3 that we are buying a fruit of tomato now can buy five of the same size in October but a K1 size of onion now will only get two in the dry season, so why are we crying about markets for our crops? What beats me is that onion is actually not as difficult to grow as tomato. It is attacked by very few diseases and pests as compared to tomato. It is not a heavy feeder in terms of fertilizer as compared to tomato. Therefore, as we think of the green revolution in this country and making Zambia a food basket for the region, let's think of including onion in our enterprises. Honestly, we can't be importing onions from Tanzania and South Africa when we have so much land and water. Furthermore, I feel as though I haven't done my job to learn that we are importing garlic from far flung places such as China which has so many mouths to feed as well as rice from Thailand. In the next article, I will endeavor to elucidate how growing and branding of rice can create jobs in this country.

Impact of rice branding on job creation

To most Zambians, what comes to mind when asked whether one has eaten is nshima (*ubwali, pulp, dzsaza*) and chicken or meat? Even kids when asked whether they have eaten, they will answer in the negative that they just had eaten rice and chicken (*tulelolela ubwali, tatulalya*! meaning they are waiting for nshima, they haven't eaten yet). For the sake of the visitors, both tourists and investors that could have just come into our lovely country and have laid their eyes on this book, *ubwali* is a thick porridge or paste derived from maize meal. In east Africa, they call it *ugali*, I don't know in South Africa though I have heard some calling it as 'pulp' (which is a street language here in Zambia or *akamege*).

Rice is believed to be a plant that is native to India and Thailand in Asia. It is the second most grown cereal in the world and second only to wheat. This crop grows well in hot humid conditions. In Zambia, we have conditions that favour the growing of rice. Until recently, the only varieties that were available in the country could only be grown in very wet conditions i.e. in environments that favors water logging and have relatively flat terrain with suitable clayey soils. It is the reason that we find a lot of rice cultivation activities in Western and North Western provinces because of the Zambezi plains. Actually, the rice belt stretches from Zambezi-Chavuma to Lukulu, Mongu, Senanga and parts of Sesheke. In Luapula province, it is grown along the Luapula-Bangweulu-Mweru stretch. To be specific, it is areas around Chembe, Chienge's Maoma, Lambwechomba and Lambwechikwama up to Kaputa in Northern Province. We also have the Chambeshi basin in Northern as well as the Luangwa that goes up to Chama and Mfuwe and Lundazi districts in Mchinga and Eastern provinces. This is not to say that it is not grown elsewhere. I was surprised the other time that I found some small-scale farmers growing it in smaller stretches along the Mulungushi River in Central province and I suspect there should be slightly larger scale production along the Lukanga swamps. This indeed is a sign that our farmers are receiving well the information of diversifying but what could be worrying is the speed and area put under such crops. For the sake of those that fear water like me, there have been varieties that have just been developed which can be planted on the upland called, NERICA (New Rice for Africa). I should guess even those varieties grown in paddy rice can do well on the upland; my thinking is that they grow them in flooded areas to suppress weeds. I will need to find out from my good friend who is a researcher in Mongu under ZARI.

In this book, I will not very much dwell on how to grow rice because what we want to establish is how easy this commodity is to brand. However, for those that might be inspired and want more information on how to grow rice, they can get hold of a good friend of mine mentioned earlier who happens to be a rice research officer (specialist) at ZARI based in Mongu. Though I know a bit about it, I am hydrophobic (the fear of water), so I did not take much interest in knowing a lot about the detailed agronomic practices about this crop. I have always liked to work on the rice value chain when it has been harvested. I fear leeches that are sometimes found in paddy fields.

Globalization is having an effect on the feeding patterns of Zambians in that a good number of people are able to eat rice as their main meal. Apart from the areas that are producing rice in Zambia, most of our rice that we are buying on the market and supermarkets are coming from far away countries such as Thailand, Tanzania and Malawi. Do we know how many jobs we create in Thailand every time we eat Thai rice? Do we know how many children we deny to go to school every time we ignore to buy our own Mongu, Chama, Kaputa rice? The easy answer many of us would give is that the other products are well packaged and labeled. Well, let us talk about the conditions our rice is marketed in. Just to show how our farmers are trying to diversify, I will endeavor to bring our some of the pertinent issues about the conditions in which the farmers and traders of Chienge and Mongu goes through to produce and market their rice.

Chienge and Kaputa rice: There is always been a very big debate as to which area grows more rice than the other between Chienge and Kaputa districts. This is because you will find that more rice is being processed in Kaputa than one would find in Chienge. Agricultural

Extension Officers usually would argue that Chienge grows more rice than Kaputa but Kaputa has better processing infrastructure. The growing belt for rice is near Kaputa so a lot of farmers take their rice to Kaputa not only for processing but also an as distribution route through Kasama to Lusaka or copperbelt. I used to agree with this line of thinking because the road from the production areas (Lambwechomba) to Chienge boma where there are processing facilities is not there, even with a motorbike; one has to be careful otherwise he might just find himself below the famous Ntamba Mweru hills. The point am trying to drive home is that you can only talk about value addition and branding with relatively good infrastructure such as roads, electricity and others. There is literally no or less value addition for rice coming from Lambwechomba yet they produce more than Kaputa. However, some few residents with entrepreneurial minds have managed to buy diesel propelled machines and are milling on a small-scale though the quality is not as good as compared to that in Kaputa. Most of the rice, however, is milled and packed in those big 90kg bags. When they are packing, they use small sticks or pestles to pound it so that a 90kg bag will weigh around 150kg however, this compromises the quality as the grains are broken in the process. The farmers want to maximize on transportation as usually truckers charge per bag and not per kg as it was. This make the product to fetch relatively lower prices than it would have, had it been nicely packaged.

Mongu Rice: Mongu rice is produce along the stretch from Lukulu to Senanga including parts of Kalabo, and the variety cultivated is called *supa* rice. It will be good to note that farmers along this production area grows more rice than maize, yet the government through the FISP provides more support to maize than rice and the area has comparative advantage in rice production as compared to maize. However, these farmers have challenges of where to get pure seed, otherwise, they have been using recycled seed for ages. I was amazed to see demo plots of NERICA variety in some field days I attended but never saw any spa rice. Mongu rice is well known almost everywhere in the country, they say it has a nice aroma and tastes good. We are yet to know which areas amongst the three (Lukulu, Mongu and Senanga) grows more rice. Mongu has relatively better facilities for processing in that there are a lot of SME that are polishing it, such as DMDC, Sefula Farmers Group and others. We also have companies like APG Milling that have set processing units there and National Milling have established a depot where they bulk and transport it to Lusaka for processing. The farmers and retailers in Mongu do polish and package the rice in 5kg and 10kg bags labeled it as Mongu Supa rice. These products are found even in supermarkets like Spar, Shoprite and Pick n Pay. There are hundreds of retailers in Mongu, Lukulu and Senanga that sell rice but have no fields of rice. You will also find a lot of middlemen selling paddy rice to millers and traders. Well, now that we have contrasted two rice production belts, lets continue with the nitty gritty of branding rice for job creation.

Branding helps to tap into the mainstream markets: with the ever-changing weather patterns in our country, we should embrace new eating habits and adjust our cultures. So, what is a brand? A brand can be defined as 'a collection of images and ideas representing a product or service. It refers to the descriptive verbal attributes and concrete symbols of a product or service. A brand gives an identity to a product and eventually the company which released it. The brand tries to keep the consumer stick to the brand and refuse to use other products'. Like many other Zambians, I am a strong believer that availability of a market for a product will stimulate its production. I believe that the market for rice in Zambia is enormous, it is the reason why we are having products coming from as far away places as Thailand, India, Tanzania and South Africa to just mention but a few.

However, you will have to write a full thesis to convince me that Thai rice is better than our Mongu rice or Chama even Kaputa, but then why is it that Thai rice seems to be more popular than our own rice? A lot of answers given to this question are that it is cheaper than our rice. How is rice that is transported several thousand miles away cheaper than rice which is produced only six hundred kilometers away from Lusaka? When we talk about price, we are already talking of Thai rice carrying the brand image of being cheap. Why then can't we make our Mongu rice affordable? Am sure this will be a topic of its own and we shall discuss it later with the help of some marketing principles. There are several attributes that one can use to brand a product or service, some of which are quality, price, appearance, lead time (delivery), availability, service, taste, durability, ease of use and so on.

The beauty about creating a good brand image through various value adding activities is that you will make the product/service sell easily and if a product is easily sold, then it will have available market. Every time we sell/buy a finished product, we are helping create or sustain several jobs along the value chain for that particular product. Take for instance, every time we buy Mongu rice or Kaputa rice in Pick n Pay, we are sustaining the job of the shop vendor and the young saleslady in that shop, we are also doing the same to the transporter who delivered the rice to Lusaka, the trader who went and bought rice from the farmers and the farmers themselves, the processor who helped polish that rice, the trader at the market who sold the grain bags to the rice trader, the seasonal worker who helped till the land for the farmer to plant, the input suppliers who sold the seed and fertilizer to the farmer, and the chain goes on. I have tried by all means to avoid numbers as some people may argue that business without numbers is nothing. Indeed, if we are talking business in agriculture, we need to talk about the volumes that are produced and sold, the quality of the produce, the pricing and hopefully the improvement in the livelihoods of the number of farmers we are impacting.

However, my interest in this story is that we should try to create good brand images of products through various value addition activities in order to create jobs. I have an example that I want to share with you. In all efforts that we are trying to put by helping the smallholder farmers, let's by all means ensure that we try to link them to mainstream markets if these activities are to be sustained. And by helping them sell branded products, we are helping them sustain their markets. If one goes into Shoprite stores, besides other rice brands that are there, one will find National milling's Mongu rice. If you go into Pick n Pay, Spar and Melisa supermarkets one will find nicely packaged and labeled Mongu rice by various suppliers. I will not talk about National Milling lest you say; it is a corporate entity and hence has the capacity to brand their products. Let me give you an example of SNV and Concern Worldwide. These two non-profit making organizations are helping farmers in Western province to tap into the mainstream markets using the model SNV is calling Inclusive Business. Concern Worldwide is supporting the farmers through the district farmer association by offering extension services to the farmers and they have provided support to establish a polishing plant for the association.

On the other hand, SNV is working with local processors to build their capacities through several fronts. They provide trainings and coaching in business management which involves quality processing, financial management and marketing. Where they have seen that the management capacity has been enhanced but the firm is probably lacking financial capacity to recapitalize their businesses, they provide finance brokering services and help the firms get loans or grants. They have gone a step further to help link these processors to markets, an example being the linkage facilitated among DMDC (a local social enterprise) with Pick n Pay, Melisa and Spar supermarkets. They have also facilitated a lot of trainings in rice production and with the help of ZARI; they have published a rice growing training manual to help with extension. In 2010, an association/movement called Zambia Rice Federation was formed and registered to spearhead the rice value chain in the country.

The morality behind this story is that there are a lot of organizations, programs and projects that are involved in trying to increase rural household incomes through improved productivity. My concern is that even if we offer support to improve productivity, it will not be sustainable if we do not try to link the farmers to mainstream markets or to realize the potential markets. This reminds me of the article that was published in my column in Times of Zambia where I jokingly talked about my friend Katungu who always thinks a market is only when he is able to sell to the Queen of England. If we link the small-scale farmers to mainstream markets, then we shall be able to sell to the Queen of England indirectly. Mind you, supermarkets like Game and Pick n Pay are international. Linking of smallholder farmers to mainstream markets can be done in several models. Just those minor efforts by those organizations and others have made a difference between Chienge and Mongu rice. The biggest difference being that even if Western province has the worst roads in the country, the production areas for rice are relatively near to the processing plants and the main road, unlike Lambwechomba in Chienge district. With the help of basic infrastructure, rice is relatively easy to brand.

There is another model being used by Techno Serve Inc. in linking farmers to mainstream markets. They were collaborating with Cargill in that some farmers that are working with Cargill in cotton value chain are being pre-financed in soybean production. The company thereafter buys all the soybeans that the farmers produce. The idea is that farmers should be able to effectively rotate cotton, soybeans and maize thereby improving the productivity of soil. Being a trading company, they are able to access international soybean markets unlike the smallholder farmers who do not have the voice or capacity. I believe we have seen here the link; tapping into mainstream markets can create and sustain jobs in the agricultural sector. However, I have exclusively decided to leave out the role of the policy formulators (government) because that will be discussed later in this book although I have tried to introduce it in earlier chapters. One missing link in this puzzle was where to source the seed by the farmers. With my negotiations when I was working with SNV in Mongu, we started an activity with ZARI to purify the spa rice variety. This will be discussed in the next article where we would show the importance of pure seed in enhancing productivity for any crop.

Investing in seed purification

In 2010 while working for SNV in western province on the rice value chain, I observed that the most preferred variety of rice in that part of Zambia was Supa rice sometimes called Mongu rice. It is the variety which is grown throughout the buLozi (Barotse) plains. Our drive was to improve productivity and increase household incomes through enhanced value addition services such as processing and branding. However, one striking setback noted was that the seed used was recycled year in year out – there was no source for hybrid seed. Nonetheless, the people have kept growing the crop but the definite result is that productivity can only be increased to a certain level as seed would be the limiting factor regardless of the improvements on the other factors of production. One day as we were having a cold drink at Dolphin lodge with a friend who is a rice breeder, we discussed this issue and he was agreeable to the fact that there is need to 'clean' up the variety if rice productivity was to be improved. In principle, we agreed that the cleaning services needs to be done but we needed to look for resources for this activity. It is not easy in Zambia to get financing for enterprises that are not commercialized. He further mentioned that the minimum period it would take was three to four seasons or years. That did not resonate so well with me because I had a two-year contract and was so keen to see productivity improved by the smallholder farmers that where growing the crop. The motivation was to help these resource poor farmers increase their returns from their investments in the rice value chain. We however, kept sharing notes on the stakeholders we needed to bring on board. I was so elated in 2015 when the same friend from ZARI informed me that he was traveling to Lusaka to make a presentation on the release of a pure Supa rice variety. It has since been cleaned up and we now have a pure Supa rice variety. This was a milestone for me except to state that there was one huddle left; ZARI is not involved in commercialization of seed inputs. I guess they need a serious input supply private sector to get on board to start multiplying and distributing that variety to the farmers. We have in the past ten years been talking about crop diversification but the serious crop diversification started with the current season when the government allowed farmers to get inputs of their choice unlike what has been traditionally been happening. I would frown to see farmers using urea for top dressing in waterlogged rice fields instead of the ammonium nitrate just because that is the fertilizer provided by the government.

The second constraint I want to see worked on is the white beans. I was privileged again to have worked in Luapula and parts of Northern provinces. There is a bean growing belt which stretches from Mansa-Mwense into Kawambwa's Musungu area up to Luwingu and beyond. Farmers in that belt grow a lot of beans and the variety they like growing most is the white beans. As a matter of fact, it is that type of beans I enjoy eating as well in preference to other varieties on the market. Just like the spa rice, this variety has no hybrid seed and during the time I interacted with the farmers, their yields were relatively low because of impurities in addition to other factors. If I remember my agronomy well, beans just like groundnuts are open pollinated. However, the productivity vigour decreases with the way the seed has been stored. Like I have shared earlier on this book, Luapula province is amongst the few provinces that do not ask for relief food because of their production systems. ZARI Mongu have indeed added to success of rice by cleaning up of this variety called Supa rice. I am urging the legume team to emulate what their colleagues have achieved by cleaning up this bean variety and sell the rights to private input suppliers to start commercializing that variety. I am alive to the fact that in the country, we only have Pannar seed that is selling a pure bean hybrid. We need to add to the varieties just like we have done with soybeans; we have SeedCo with about three varieties or so; MRI with two varieties, ZamSeed with two. With the rain patterns shifting and favouring Luapula, Northern, Muchinga and Northwestern, I can smell an opportunity in rice and bean growing besides maize. Currently, we are not self-sufficient in rice production as we are still importing from Thailand and the countries from that region. With the meat prices becoming almost unaffordable by the majority of the population, beans remain the best substitute for this market. Mind you, meat is yet to become even dearer as the global population soars. Growing beans, is the best substitute for the meat protein. In other cultures, bean is favoured more than meat. For instance, the Namwaanga's say *wiponya chooli uponye umwana* (they love means so much). Meaning don't drop the container with beans instead drop the child. This just shows how much they value beans. Suffice to say the yields of beans have remained stagnant at 500kg/ha for a long time and some problems mentioned in this article contributes to this low yield. This is a big opportunity for those that studied agriculture like myself to offer agronomic advice to farmers at a fee like I do to tomato farmers in my spare time. Don't complain of not having jobs in Zambia; jobs are everywhere. What is important is how you apply yourself. Agriculture is a potentially live industry in Zambia with so many opportunities.

A dead yet viable subsector

A lot of you are already wondering about what I mean by saying a dead yet viable industry. We all know that groundnuts are readily available in Zambia. Yes, it is very true that groundnuts are readily available in Zambia. Actually, it is the second most grown crop after maize. It is grown in all the ten provinces of Zambia with Eastern Province being the leading producer in the country. In the world, China produces the most peanuts at about forty-five percent followed by India at eighteen percent. In Africa, Nigeria is the largest producer followed by Sudan, Chad and Senegal in that order. There are only three countries from Africa in the top ten, with Nigeria at five percent (two percentage points less than USA). Although China is the largest world producer, it is one of the major importers too followed by the European Union (EU).

In Zambia groundnuts is mostly grown by women as it is considered to be a woman's crop. It is grown on small plots ranging in sizes from plots as small as quarter hectare (0.25ha) to one hectare. The crop is usually grown with no inputs apart from the seed and it is mostly planted on newly opened up land. I remember, way back in the village when grandmother would ensure that she plants groundnuts before any seed. She would not even allow us (boys) to help in planting as there was a belief that groundnuts planted by men do not perform well. This crop forms a very important component of nutrition at household level especially in rural communities. In the village, they rarely slaughter livestock though they have chickens. We could have meat once in a month while every day's meals comprised of the greens (vegetables) and groundnuts were important to make the relish tasty. As a matter of fact, my best relish up to now is pumpkin leaves with pounded groundnuts paste. The groundnuts could sometimes be pound into a paste (*icimponde*) and we could eat with nshima. They could add this nutritious stuff to literally everything from dried meat, sweet potatoes, *mankolobwe* (dried pumpkins), dried fish especially *kapenta*, porridge, name them. One would not believe me that at some point Zambia used to export groundnuts to the EU as well as countries in the region? Yes, she did, the time of National Agricultural Marketing Board (NAMBoard)! Actually, she stopped exporting in the early 1990s when it was discovered that the quality of our products/nuts were going down; they had too much levels of Aflatoxins. Aflatoxins (mold) is a fungus which is believed to cause cancer when consumed in large amounts and can also cause stunted growth in children (I hope it's not the reason why most of us Zambians are short). Production of groundnuts also dwindled with reduction in market opportunities. Some of the firms that were processing the peanuts were also dubiously and senselessly privatized in the mid-90s, it was plunder of the waste order that was happening at the hands of the educated yet greedy Zambians.

Of late the world market for groundnuts has been on the increase again. Even here in Zambia, we are failing to meet the demand for quality whole nuts. There are four to six companies that process groundnuts into peanut butter. These companies have been demanding for groundnuts to be used in their factories but the demand outmatches supply. Why is it like this? There are so many reasons why our production has failed to pick up and meet demand. Let me share with you three major problems besides so many; firstly, we are a country that has put more emphasis and offer advice in maize production. There has just been too much emphasis on maize production. The second reason is that our productivity (yield per unit area) is far much lower. The average national yield per hectare is 0.44t/ha. This is less than USA whose average yield is 3.80t/ha and even far less than South Africa at 2.33t/ha. Our yields are far much less than even Sudan which is mostly a desert. One of the contributing factors to the low productivity is the reason we want to ponder on so much in this article – non-availability of improved seed.

Can you imagine that the crop which is the second most grown after maize has no private company that produces hybrid seed? Talk about all the seed companies around, not even a single company is producing any seed. Improved seed ends at research and no company has come up to this challenge. All the farmers that grow groundnuts are either using grain or recycled seed to plant for their commercial crop. Zambia Agricultural Research Institute (ZARI) through Msekera and GART have done their best to release high yielding and oil containing varieties like MGV4, MGV5, Chishango and many other varieties, but surprisingly no seed company has come on board to multiply these varieties for commercial sale. Sad enough, no commercial farmer grows groundnuts; the largest area I have known to be put under groundnut production is six (6) hectares by some emergent farmer. If you calculate the gross margins for groundnuts and compare it with that of maize (smallholder farmers), it is more profitable to grow peanuts (*nshaba*) than maize, but because of the brain washing by the government, we all go for maize even if it means getting our money in February of the following year for a crop supplied in July. For groundnuts, one is paid cash on delivery and no promissory notes besides it is a very important component of our nutrition. For your information, groundnuts are more nutritious than maize and I may not be wrong to insinuate to the high populations found in Eastern and Southern provinces which are the two leading production areas.

The other let down has been with our laws and policies. According to SCCI, a farmer growing seed should have a minimum of five hectares for one to qualify to produce seed. With the crop predominantly grown by the smallholder farmers, most of the seed that they produce are classified as QDS (quality declared seed); breeders can amplify what this means. We need to repeal our laws and policies on seed multiplication because farmers of 1900s are not like farmers of 2000s. They have been trained on seed production, so the area shouldn't be a limiting factor for one to produce seed as long as SCCI inspects it and meets the basic requirements. The government should facilitate capacity building of SCCI so that they can effectively monitor groundnut seed multiplication. What is most annoying being that the same government that refuses to recognize the smallholder seed grower's floats tenders to buy groundnut seed from the same farmers as was the case in 2012 planting season.

What the suppliers of those seeds used to do (I presume) was that they were buying grain from Soweto market and sorting it before supplying it to FISP as seed. Can the government ask their suppliers from where they got the seed that was distributed to farmers? Some small seed growers like the women groups in Eastern Province that works through SMEs involved in seed multiplication are disadvantaged. We need to quickly put our house in order as regards the groundnut industry. I am asking the minister of agriculture to quickly form a task force on the groundnut industry which will comprise members from SCCI, ZARI, ZNFU, FRA, MAL, Processors, traders, farmers, transporters, financiers and I am offering myself to sit on this task force free of charge. We have a very big opportunity in the groundnut industry as it also helps to sink the atmospheric nitrogen (fix nitrogen in the soil) and offering an alternative crop for rotation. We need to take this with the seriousness it deserves as groundnuts forms a very important part of Zambia's food security. In this same indaba, we need to discuss the issues to do with aflatoxins as this is a hindrance to exporting our nuts as well as a health risk factor to the population.

Aflatoxins in groundnuts

I should state from the onset that I'm not an expert in this area but I will try to give you an idea and try to apply it in business sense. If you want to get better details, you are advised to contact someone from the Food Science department of the School of the Agriculture Sciences at UNZA or any Food Technologists near you or any crop pathologist. Aflatoxin (*Aspergillus flavus*) in groundnuts is a trade and health constraint for development of the subsector globally.

Have you ever thought of why you pay more for a packet of groundnuts bought from Shoprite than the same pack size at Soweto market? Many of you would say because of how it is packaged, yes to some extent you may be right. However, there is more than just packaging for the value of the groundnuts you will buy from Shoprite[9].

---

9 Shoprite is chain store supermarket from South Africa

This value is perceived because most of the times you will get the same quality although chances of buying bad quality nut at Soweto are higher than that from Shoprite. Bad quality comes due to contamination that occurs due to poor handling. The contamination we are talking about is due to Aflatoxins. Aflatoxin is a carcinogen which can cause liver cancer if consumed in large quantities. There has also been a link between Aflatoxin exposure and stunted growth; health practitioners may provide more details on this topic. Aflatoxins are small molecules toxic to both humans and animals. They are produced by two fungi; *Aspergillus flavus* and *Aspergillus parasiticus*. They are found in groundnuts and maize but more prominent in groundnuts. This fungus comes about due to substandard conditions. The mold produces aflatoxins. These conditions are caused by insufficient storage facilities which lets in moisture and humidity that creates mold.

There are basically six entry points for aflatoxin in the production of groundnuts from pre-harvest to storage. Pre-harvest entry points are at seed stage and bad on-farm practices while post-harvest entry points are due to shelling practices such as adding water to soften the shells when hand shelling, buying practices (buying nuts in an unhygienic environment as the case is currently at Soweto *matipa* this time of the year – rainy season), poor in-fielding drying methods and poor storage facilities as earlier mentioned. Pre-harvest infection is significant in the semi-arid tropics, especially when end-of-season drought occurs. Poor post-harvest conditions in warm humid area, and bad harvesting and storage practices lead to rapid development of the fungi and higher levels of toxins. This is especially true in developing countries like Zambia where preventive measures are frequently ignored.

Consumption of Aflatoxins by human beings can lead to liver cancer as earlier stated. A person's chances of contracting cancer are compounded significantly if he/she carries the hepatitis B virus (the virus that causes jaundice). Like I mentioned in the earlier article, contaminated groundnuts and groundnut products cannot reach lucrative international markets. This is the reason why exports from Asia and Africa have declined due to the stringent quality requirements of importing countries in the developed world such as those from the

EU and America. The aflatoxin levels are measured in parts per billion (*ppb*) and if I can remember well, the threshold to export to EU is 4*ppb*. Health issues in the developed world have been largely addressed, but elsewhere the situation is very different. Some studies conducted by ICRISAT in one of the Asian countries revealed aflatoxin levels as high as 40 time's permissible limits.

This challenge is not insurmountable; we can overcome it if we all put our heads together. Other countries have moved steps ahead in the region and one of such is South Africa. They have set up sophisticated laboratories with equipment which is able to detect this menace to the standards required by EU. There have been organizations such as the USAID that are helping developing countries in Africa to overcome this problem. However, it is very difficult to fight this constraint if the private sector is not involved. In Zambia, a laboratory is being installed in Lusaka at Mt Makulu that will be used to measure the aflatoxin levels in the nuts that we produce. The problem we have in Zambia is the bad attitude; if it does not affect you immediately, we care less but when it reaches a level of a pandemic, that's when you hear workshops being called even in taverns. I therefore, urge all those that are involved in the groundnut industry including you and I the consumers to come together. I complained that there is no private seed company involved in groundnut seed multiplication, many of you thought I was politicking because politics is what we know best. If we can't fight this constraint, where are we going to sell our groundnuts and groundnut products? Our population is just too small to absorb all the nuts if our productivity was to even increase from 0.44t/ha to even a paltry 1t/ha. We should not always rely on the informal market from Congo because the Congolese are slowly waking up and sooner than later, they will be self-sufficient in most commodities as soon as they stop fighting.

Our biggest opportunity for agricultural products is the EU because there are certain products that can't grow in that part of the world besides they have limited land left for agriculture. We should start strategizing how we shall be the food basket of Africa in 30 years' time when there will be more poverty than there is today. We have the land, the water, the human resource but why are we still under producing?

If it is lack of political will, let's compel our leaders to buy into what we want to be. Who thought MMD will be out of power today? If you did it to remove that unresponsive party from power, let's ask the current political leadership to buy into what we want to achieve. Once that is done, lets solve the problem of having pops in our groundnut production activities.

Losing money due to 'pops'

Stopping at some roadside makeshift on my way from Mkushi in 2015, I decided to buy a tin (5kg) of fresh peanuts for my family. I normally prefer the large ones locally called *Chalimbana* to the small popular *Solontoni*. At face value the groundnuts looked well filled and ready. Little did I know that I was dealing with seasoned business ladies who know how to package their merchandise to deceive their-would be customers; business crooks in short. They had put groundnuts that were not mature with so many pops in the middle of the tin. When I reached home, I looked as though I was buying in the night as my wife rebuked me for buying such poor quality nuts. This was after I was commended for buying good nuts a week earlier when I went to Chipata, but how was I supposed to know?

The question that seeks answers is 'what causes some groundnuts to have pops on the same plant'? In Zambia groundnuts are the second most grown crop to maize by the smallholder and emerging commercial farmers. Predominantly, it is grown by the women farmers on small plots or farms ranging from quarter hectare to a hectare at most. There are so many things that cause groundnuts to have a lot of pops; and top on the list is the low soil pH or high acidity. Groundnuts are so sensitive to soil acidity and this is because calcium which is one of the key nutrient in groundnut production is usually absent or unavailable in soils with low pH. To effectively produce a good crop one need to check the soil pH and rectify it to near neutral. This can be done by applying lime to the soil so at to increase the soil pH. Secondly, calcium is a main component of the 'hook' that grows from the branches into the soil on which the pod forms. Low available calcium will have a negative effect on the quality of shell and grain in general. Groundnuts also need soils that are 'friable' in that compacted soils are not ideal for groundnut production because the hook on which the pod forms fail to penetrate the ground. I remember as a boy, my grandmother would chase me or prevent me to access her groundnut fields while wearing shoes. Though this was more of a cultural belief, it has some explanation with soil compaction. Her argument was that if I go in a groundnut fields with shoes, the groundnuts will not fill. I think the scientific explanation behind that is that moving in the groundnut field with shoes on might contribute to the soil compaction.

I normally give the women farmers in Zambia a big hand in terms of weed management in groundnuts; if only they can keep their maize fields as clean as they do in their groundnuts fields, we could be talking of something else in terms of productivity. Suffice to say that the weeding is mostly done by hand but there are herbicides that can be applied to control weeds in groundnuts just like any other crop. Lucky enough, groundnuts compete favourably with weeds as they quickly canopy off the weeds except for upright groundnuts. The third factor I have seen which leads to poor harvest of groundnuts is that this crop is highly affected by both fungal and bacterial diseases. I have walked in a groundnut field and observed about more than two diseases affecting the leaves and when you tell the owner, she will answer that that is how the crop looks and not necessarily being a disease. This is a huge opportunity for agronomists to provide extension service. The truth of the matter is that when most of the leaves are affected by diseases, it reduces the surface area on which the crop manufactures plant food which is supposed to be translocated to the pods to fill up the grain hence, the pops. Smallholder farmers in Zambia lose about twenty percent of yield due to plants having affected leaves and this leads the plant to dying prematurely before it completes the grain filling process. What is annoying is that there are so many fungicides on the market that can easily control most of these leaf diseases. When you want to apply, such fungicides ask for those that have a combination of a strobys and a triazole in that the stroby will provide that greening effect which is very important for complete pod filling and stress management. Companies like BASF has produced fungicides that have an excellent greening effect which has been branded as AgCelence effect. One of such product is called Bellis and it is good for groundnut production as well as other crops such as vegetables.

The other factor that I have noted which makes us not attain at least more than a metric ton per hectare is that we rarely apply any fertilization in groundnuts. Although we know that groundnuts are nitrogen fixers they still need other elements other than nitrogen which they can't get from the atmosphere such as calcium. It would be good to give an initial boost of basal fertilizers with less nitrogen in it at reduced rates if we inoculated our seed. Additionally, we need to be inoculating the seed at planting just like we do with soybeans to help the plant with fixation of nitrogen. Lastly, even though groundnuts seed can be retained, we need to be using improved varieties, probably every after a season because home kept seed can lead to reduced yields depending on how well it was kept and it might harbor diseases. We have heard of Aflatoxins in groundnuts and this can come through seeds as well. So next time we buy those groundnuts at the roadside with so many pops lets interview the owner to find out whether she knows the reasons why her merchandise is like that. Root health are also very important in crop production and it's important that we discuss it in this book. Generally, groundnut prices have been profitably stable for over five years now. Only if farmers can improve their productivity to be attaining yields as high as a ton and half, many farmers would switch from the maize production to groundnut production.

The foundation for effective crop production

After publishing an article on soil acidity in one of the daily newspapers in 2013, I received a lot of comments from my readers. In our discussions, I picked up a few points that they appreciated the way I used to write my articles in simple English. Indeed, my audience is the small-scale farmers that are mostly starved of agronomic information;

it is the reason I restrain myself so much to delve into scientific details, rest I could confuse my clients. That is the style I adopted even when writing this book. However, I got the feeling that most of the readers do not appreciate the importance of roots to crop production and later on productivity. I have added the article on soil acidity in this book too, though it is not as detailed as soil scientist like myself would put it in writing a scientific paper.

Do you know that a farmer is the only worker in this world who is supposed to know all other skills? An accountant only knows about numbers, an engineer knows only about engineering and basic accounting and so forth, but the farmer is the only person who has to know everything; accounting, agronomy, engineering, pathology, mechanics, human resource, and every other trade that you may think of in business. Farming is a science as complicated as a doctor that treats babies; you don't speak to plants for you to know what they are lacking but you have to use your eyes to deduce exactly what is wrong with the plant to institute remedial measures. A child's good up bringing starts from the day of being conceived while productivity of a crop starts from land preparation and choice of genetic material of the seed. Once all these are done and the crop germinates, that will be like conceiving. If the roots are not well developed, no matter how much nutrients you apply to the crop, it will only be able to get a fraction of what it is supposed to. This is because the 'mouth' of a plant is based on the roots while the nostrils are on the leaves – what a face!

The roots take up water from the soil as well as nutrients. It also releases some exudates to maintain osmotic pressure as it gets nutrients. If you get a leaves of a wilting plant and deep them in water with a view to resuscitating it; you are just wasting your time or may be called to be insane. It is imperative then that we have to ensure that the roots of our tomato, cabbage, maize, soybean, mango, peas, carrots, name them are well developed before we even think of applying that fertilizer. How then can we ensure that the root development is at its potential? Firstly, ensure that your land is well prepared. By saying so, I don't only mean that it is well cultivated. It has to start by ensuring that the soil has the right pH for the type of crop we want to grow. If you are growing soybeans for instance, ensure that your soil pH is always above 5.5 because the more acidic the soil is, the less available will be phosphorous; an element which is key in soybean production. Many crops do not develop good root system in acidic soils; there are very few plants that can tolerate acidic soils such as pine trees. Additionally, ensure that you do not have a hard pan within 30cm of soil depth. The crop will use a lot of energy to break through this hard pan otherwise, the roots development will be within the surface and this will negatively impact on your crop accessing deep lying nutrients. It is the reason our good friends and partners from CFU encourage ripping instead of ploughing. I am a fan of ripping as well. Do you know that a crop with a well-developed root system is an assurance of at least thirty percent of the potential yield? This is because a well-developed root system will increase the surface area with which a crop/plant is able to scavenge for nutrients and water in addition to anchoring the plant firmly in the soil even after it has dried.

There is normally this bad habit with our farmers especially those that don't want to use herbicides of using a plough or cultivator to weed their fields – I don't like that habit, why? As you are passing through with your plough in the inter-row, you tend to cut and injure some roots. You are not only endangering your plants from being attacked by diseases because those wounds will be openings for pathogen infestation but the crop will also use part of its energy in the healing of those wounds. I know my grandmother just like yours, used to tell us that we need to open the soil for aeration. It was an important point for perennial crops and not crops that only needs maximum three months to mature. Soil aeration has to be done at land preparation and that is enough unless you plant your crops in water logged areas. You also need to protect your crops from nematodes with nematicides. However, this will be a topic for another day. Therefore, the importance of good root development in crop production is non debatable. Those roots will also contribute to soil organic matter after harvesting your crop, so what is soil organic matter and of what importance is it to the soil and crop?

Soil organic matter

From the time I started publishing agronomy articles in the newspapers, I had received numerous calls for me to put the articles together in form of a book. A lot of my readers had pleaded with me to do just that. So I started writing this book in 2016 and my target was to publish it in 2017; unfortunately, I lost my laptop on which I was putting those articles together later in 2017 and I changed the period in which to publish the book to late 2018. I was foolish not to have been saving the book on my external drive. However, a lot of people have written to complain that 2018 is too far away. Therefore, I have started putting together the articles and God willing, we should have the book ready towards the end of the year or early next year. Lee Lacocca once said, 'you can have brilliant ideas, but if you can't get them across, your ideas won't get you anywhere.' I am therefore, inviting agribusiness institutions and companies like NWK, AFGRI, BASF, Syngenta, Amiran, ZNFU, CFU, Cargill, CropChem, ZamBeef, Emman, and many others that wish to have a chapter in my book about their products and services to get in touch with me.

Not long ago the agriculture sector commemorated the first soils day; this is to appreciate the importance of soil in agriculture. Soil can be interpreted differently by different people. For instance, soil can mean a surface which supports foundations to a civil engineer while with a plumber; it can mean a nuisance which blocks the sewer system. However, soil means the lifeline in agriculture. Although modern crop production can be done even of media that is not soil; soil is the main media used in crop production worldwide. In this article, lets share the importance of soil organic matter; a component of the soil.

What is soil organic matter? Many of us have different and weird explanation of soil organic matter; some think that when you have crop residue in the field then you have soil organic matter. That is not true because not all crop residues left in the field will be converted into soil organic matter, though the more residue you leave in the field the higher the potential for organic matter build up. There are some physical and chemical transformation which takes place to convert crop residue and other material into soil organic matter. It is very difficult to differentiate soil organic matter from the soil; when you have leaves in the soil, it does not mean that it is organic matter but it can be a source of organic matter. Suppose the leaves are later burnt! What is then the importance of soil organic matter? There are so many importance of this soil component to crop production. You have heard several times emphasis from agronomists, scientists and agriculturalists the importance of organic matter to crop production. One of the most important factors is that it helps to hold water in the soil. With the current weather conditions that we have in the country, soil organic matter in the soil is very important to act as a reservoir for water holding capacity for the soil; it may just be your savior with these scanty rains. Secondly, organic matter is made from once living plant materials such as the leaves, stems and roots of dead plants. These materials at some point in their growth needed elements like nitrogen, phosphorus, calcium, magnesium, iron and many other elements for health growth. As the plant materials die and decompose, they release these same elements into the soil which can be taken up by our growing crops in the soil. Therefore, they save as reservoirs for plant nutrients in the soil.

Additionally, soil is made up of micro cementing elements which hold the soil particles together. When soil particles are held up so tightly together, the plant roots might not penetrate it to seek nutrients deposited deep down the soil profile. They form a condition known as soil compaction. With the presence of organic matter in the soil, it makes or loosens up the soil; a term called being friable. The importance of the soil to have these characteristics is that if a soil is compacted, roots will need to utilize so much energy for them to penetrate the soil. No farmer would want the crop to expend so much energy on root formation at the expense of the harvestable part like the cob. This will mean the size of the cob; pod or leaves will be smaller as most of the energy was used to develop the roots. Therefore, when CFU experts tells you to be ripping your fields at least once every three seasons, it's not that they want to increase your production costs but they want your crop to get the most benefit. The process of the decomposition of dead plant material to release essential plant elements is called mineralization.

The agents of mineralization are small living microorganisms found in your soil. These convert soil organic matter to humic form, which is a very important component of soil. Therefore, if you have a tendency of burning your crop residues every after your harvest, then you are reducing the population of these helpful microorganisms in your soil; implying that the rate of soil organic matter conversion will be compromised. There are so many other importance of soil organic matter to your crop production but for now, let me end here. The bulk of the information about soil organic matter is too scientific and it is not my idea to make you soil scientists but just to allow you appreciate the basics. Remember, our main drive is to make good decisions in our various agribusiness enterprises. However, we need to discuss soil acidity in relation to crop production.

Soil acidity and liming

As promised, we will endeavor to discuss soil acidity in this book in simple and basic terms but you can read further in agronomy or soil science books. Many of us wonders what soil acidity is? In a layman's language, soil acidity is a measure of the concentration of hydrogen ions (H^+) in soil solution. It is sometimes confusing if we want to state it in scientific terms because it is measured from a scale of zero to fourteen. The lower the number on the pH scale the higher the concentration of hydrogen ions hence, the acidity. On the other end, it can either be alkaline in worse off cases; have sodic soils. A soil that is sodic means that soil pH is high and has high concentrations of sodium ions (Na^+) too. It is very expensive to make a soil that is sodic productive than one which is acidic. The process of making a sodic soil productive again is called reclaiming the soil. Luckily, most of the activities we do to the land promote in crop production promotes soil acidity than alkalinity.

What we all need to understand is that soil acidity is becoming a big problem in our Zambian agriculture enterprises today than it was fifty years ago. I can foresee this to be a huge problem especially as most land become productive and people will have not opportunity to fallow their fields. This will be more problematic with smallholder farmers that do not lime their fields. This if not checked will highly affect crop production and food security in general. This is because our land is becoming scarce and people are now practicing more permanent cultivation unlike a couple of years back when one had an opportunity to shift their cultivation. Acidic soils create production problems by limiting the availability of some essential plant nutrients to the plant. It increases the soil solution's toxic elements such as manganese and aluminum which are readily available in soil solutions in very acidic environments. Many of us think that fertilizer, water and weed management are the only important things in crop production. It is the reason many of us has ever dared to take the soil samples from our field for analysis at various laboratories in the country. What soil acidity does is that it complexes the soil nutrients; meaning even when you have applied adequate amounts of fertilizer necessary for that particular crops production, it will not be available. It is like putting a sweet in your mouth without removing the plastic sheath and expecting to get the sweet taste; that is not possible.

Many of us will wonder as to what causes soil acidity? There are many factors that lead to soil becoming acidity but I will mention about five to six factors which are the major causes in Zambia. Soils become acidic if you have an acidic parent material. Allow me to assume that soil is formed from weathering of rocks coupled with rotting of organic material. If the soils have acidic bedrock at the bottom of the soil from which it was formed, it may cause soil acidity as it continues to weather or breakdown. The other factor which causes acidity is high rainfall and leaching. As it rains, it leaches certain basic elements off the profile leaving the aluminum and hydrogen ions, hence, making the soil acidic. This is the reason why you will find that soils that are in region one in Zambia (this is the region of high rainfall mostly found in the north bordering Congo D.R.) are highly acidic. A soil found in Mbala is most likely to be more acidic than one found in Magoye in the south. Organic matter decay can also cause soil acidity. In the process of breaking down it releases what are called humic acids. Thirdly, the harvesting of high yielding crops will mine the soil off the basic nutrients which may lead to soil becoming acidic. Whenever you harvest your groundnuts, you are mining the soil off elements such as magnesium, calcium and other basic elements leaving hydrogen and aluminum which are nuisance to causing soil acidity. This is one reason we don't recommend burning of fields even when you are clearing land. The other factor is that the continual application of nitrogenous fertilizers such as urea and ammonium nitrate without remedial effects can lead to acidifying of the soil. These fertilizers have an acidifying effect on the soil. In Zambia, we also have another factor which we do not consider much but it is more detrimental than even application of fertilizers; it is mining. When mines release Sulphur dioxide in the atmosphere and these come into contact with water at high temperatures, they react to form sulphuric acidic which rains in form of acidic rain and this has an acidifying effect on soils.

So, what can we consider to be an acidic soil? In agriculture, soils with pH less or below 5.0 may be considered to be slightly acidic, while those less than 4.0 on the pH scale are acidic and may have high aluminum concentrated enough to limit or stop root development. As a result, plants cannot absorb water and nutrients leading to exhibition of nutrient deficiency symptoms and stunted growth of crops or plants. Am sure you are sitting on the edge of your seat wanting to ask the question as to what can be done once the soil is acidic. This is simple; firstly, one needs to take their soil samples for testing or analysis in soil laboratory at Mt Makulu, Misamfu, UNZA or any other research station including some fertilizer companies. Once they know the level of acidity, then they can apply lime. I know that many of the farmers especially the smallholders find it a problem to apply lime using their hands. Indeed, it is difficult if you don't have an applicator and you want to apply about forty hectares of your field by hand. Of late, there are some liquid lime that are been applied through spraying making our work a bit easier though I should guess the neutralizing effect is for that particular cropping season only. Some companies are selling liquid lime which they are calling Cal-lime and many other brands. The advantage of this lime is that it works faster in that it neutralizes the acidity in the solution. The most commonly used liming material in Zambia is agricultural lime from limestone, which is the most economical and relatively easy to source and manage. The limestone is not very water-soluble, making it easy to handle. Lime or calcium carbonate's reaction with an acidic soil 'removes' acidity (H^+) on the surface of the soil particles. As lime dissolves in the soil, calcium (Ca^{2+}) moves to the surface of soil particles, replacing the acidity. The H^+ reacts with the carbonate ions (CO_3^{2-}) to form carbon dioxide (CO_2) and water (H_2O). The result is a soil that is less acidic (has a higher pH) and conducive for plant growth.

What we also need to know is that soil acidity is in two forms; there is active acidity which is the acid (H^+) in soil solution and the reserve acidity which is the acid on the adsorbed soil colloids or particles. As we neutralize the active acidity, the reserve acidity will slowly be released into the solution. This is the reason that we need to continue liming our soils atleast every three years. A word of caution is that lime is not a substitute for fertilizer as some people may think. We need to know that soil is a living thing; we need to look after our soils very well if production has to be sustainable.

The right time to weed

Farmers have been wanting to know what would be the right time for them to apply herbicides against weeds in their fields. On average over twenty percent of smallholder farmers' fields are abandoned every season because of weed infestation. This trend has even increased with the increase in the area planted per farmer as they endeavor to adopt crop diversification. On average, each small-scale farmer is cultivating over a hectare of land with maize. Unfortunately, some fields for commercial farmers also leave much to be desired in the region. I have watched with awe the level of management practiced by some big guys in agriculture. At some point, I have wondered whether these farmers are propagating weeds or they are seriously growing crops. So, when is the right time to control weeds regardless of the crop?

I guess we all agree that weeding with a hoe for a field larger than a hectare is not feasible for a farmer who wants to get the best out of his sweat. Weeds grow naturally and they have adapted so well to the environment that they always outcompete the crop. When we plant a seed, first it has to imbibe water at least fifty percent of its weight for it to start the germination process. Now suppose, as normally is the case, we planted the seed days after receiving rains. That will mean the weed seeds would have already started the process of germinating because they are already found in the soils before the crop starts to germinate. By the time the crop seed will start to germinate, the weeds will have emerged and established themselves with good root system. It is therefore imperative that weeds should be controlled before they germinate. I have always shared with the farmers that the best time to control the weeds is before they germinate or emerge. This means that farmers need to use and apply herbicides before the weeds and the crop start to germinate.

These herbicides are called pre-emergence herbicides and most of them work on the principle that they sit on the germination weeds shoot or root tips and prevent cell division. This will make the weeds to stop growing and will normally die unless if they are not germinating from the stolon. Where the spray was done effectively, crops have grown without applying any of the post emergence herbicide. The difference between pre-emergence and post emergence herbicides is that mostly pre-emergence herbicides are not selective or have partial selectivity meaning if you spray them on top of maize, they may affect the crop as well. While on the other hand, the post emergence herbicides are normally selective; they will only kill the weeds and leave the crop standing. However, you do not spray the post emergence herbicides before the weeds germinate. From my experience, the best weed management is done when you use pre-emergence herbicides. In my small *'shamba'* (field) I have been using pre-emergence herbicides and weed control has been perfect such that my neighbors who happens to be a lead farmer from a group that is practicing conservation agriculture has sometimes use my field for demonstrations to their farmers. The weed control in my field has always been perfect as compared to this neighbors' field.

Normally, commercial farmers will use a combination of pre-emergence and post emergence herbicides but the main spray program is centered on the pre-emergence and only goes in with a post emergence as spot sprays. The other advantage of using pre-emergence herbicides is that you don't allow the weeds to compete so much with your crop for the fertilizer that you apply. If you decide to use the post emergence, the weeds would have already 'stolen' the fertilizer by the time you spray and the yield would have been compromised though not as much as hand weeding. However, you should know that there are certain post emergence herbicides that are not selective. These are herbicides like Gramoxone (active is Paraquat), Touchdown and Round Up (active is Glyphosate). These will kill anything that is green and is actively growing including your maize, soybeans, and groundnuts and so on. These are normally sprayed as burn down to kill anything growing in your field before the crop germinates. I have witnessed a few farmers that have sprayed this on top of their maize and the result has been that they have 'cried like babies' because they lost their crop. It is advisable that you buy these chemicals from renowned suppliers and agro dealers. Please do not buy these chemicals from the streets otherwise; you will destroy your crops and cry like a malnourished child. They come with proper instructions of how to use and when to use them. All crops have herbicides that can help you weed your field; visit your nearest agro dealer or supplier to learn more about herbicides.

Additionally, you need to know the type of herbicides; for instance, with post emergent herbicides, we have contact and systemic. A contact herbicide is one that will only 'kill' the part of the weeds that has come in contact with the spray solution; it will not kill the roots. Examples of such type of herbicides are Basagran and Paraquat. However, on the other hand systemic herbicides will get in the whole weed system and kill the whole weed including the roots. Examples of such herbicides are Stellar star. As a farmer we also need to understand other characteristics of herbicides such as rain fastness. This simply means the period of time one need to wait to irrigate after having sprayed herbicides. The smaller the rain fastness the better the product; this is because we do not have control of when it will rain. As a business person involved in farming, we just have to use herbicides and need to know when to apply if we have to be competitive. There are certain weeds which are notorious to our crops and one of them is Striga sometimes called witch weed. How can this weed be managed?

Management of the weed called Striga

Sometime in 2015 I enjoyed an online interaction with farmers on the management of *Striga asiatica* famously called 'witch' weed. In Zambia, this weed has different names depending on the region one hails from. Some call it as *Kalowa, Kaloyi* and many others. This is one of the most notorious weeds I have ever known and it is one of the few weeds which can cause between 30 to 80 percent of yield loss and in some instances, it causes complete crop failure.

Predominantly, this weed affects 'grassy' crops such as maize, millet, and sorghum. It is called a witch because of the way it affects the crops in that it is a parasitic weed which lodges its roots into the roots of the host plant. It can be compared to tape worms in human beings. Though it has green leaves, it draws most of its plant food from the host plant. It's important to know that as crops or plants get mineral nutrients from the soil, they release what are called exudates; these are chemical compounds in order to maintain constant osmotic potential in the plant system. These exudates are the ones which stimulate the germination of this notorious weed. It is important to know the characteristics of this weed if we are to effectively manage it. This weed is an herbaceous, annual plant which grows to about 25cm high and is reproduced through seeds. One plant when mature can produce up to 30,000 seeds and more in a season. The unfortunate thing with it is that by the time we are seeing the weed emerge from the ground, it could have done the damage already. Worse off, the seeds can remain dormant in the soil up to as long as thirty years. After this weed, has logged its roots into the host plant, it starts drawing the photosynthates, minerals and water from the host plant. Finally, the crop will wither and die from deprivation. Therefore, this dispels some assurance that I got in that online discussion where some people thought that by leaving furrow some fields then one is able to minimize the infestation of this weed. What we should know is that there are also certain wild plants that are attacked by this weed; this means it is possible to find this weed growing even on virgin lands; land that has never been cultivated before.

However, we will not retract into our shells because of this weed; we will continue producing and cultivating even fields that we know have got this weed because our task is to produce food enough for the nine billion people in the world by 2030. Though we will not talk about chemical control in this article, we shall endeavor to bring out certain practices that can be done to manage this weed. Firstly, we know that it mostly preys on crops in the grass family; we need to strictly practice crop rotation. For instance, we can grow crops such as soybeans in a field we grew maize so that the exudates from soybeans may excite the germination of the weed seeds, but we know that they will not grow into the roots of the soybean plant. This means the weeds will not grow to maturity and by so doing; we shall be reducing the seed population in the soils. Additionally, we can grow hosts plants and allow the weeds seeds to germinate and chemically control them with herbicides that destroy the weeds before they reach maturity. However, this will take consented efforts because we know that each plant can produce many thousands of seed. There are also certain herbicides that can control the weeds to some extent and others can suppress it but that is beyond the scope of this article. Therefore, the next time you see the weed in your field, know the risk that your crop is exposed to and act quickly by preventing the weed to flower.

Do fertilizers destroy the soil?

In less than fifteen years from now the world population will increase to nine billion people (2030). Sad enough, when the population is still less by one billion from what it will be in fifteen years' time, the world is struggling to feed its population. We have poverty everywhere and worse off in Sub Saharan Africa and Asia. To complicate issues, agricultural land is decreasing every year due to urbanization, land degradation and other factors related to climate change. For instance, one of the best lands suitable for agriculture that we have in Lusaka has been converted into a residential area with buildings coming up every day. This is the land stretching from York farms going as far as where we have Baobab School and beyond. That is prime land good for agriculture. What hope do we have that we will be able to adequately feed ourselves in 2030 when we are struggling now?

The secret lies in technology advancements and adoption especially by our smallholder farmers in Africa. Africa is home to about eighty percent of smallholder farmers who produce about three quarters (75%) of all food consumed on the continent. Yet the productivity per unit land is less than thirty percent and the bulk of the reasons are to do with low technology adoption, bad agricultural practices, all emanating from lack access to affordable finances to invest in sector. In Zambia ninety percent of the food we eat is produced by the smallholder farmers yet they ineffectively use their productive land. In this article, we will zero-in on one technology; fertilizer usage. On several platforms and interactions, I have with farmers, they reiterate the fact that fertilizers destroy the soils. I have a divergent view from this though to some extent it could be right when done in a none sustainable manner. There is nothing on this planet which has no side effects if wrongly used. How many people have lost their lives because of over eating or eating the wrong amounts of food? Every day, we are warned on health talk shows that too much of sugar and salt can lead to us retiring early to the graves but does that mean those foodstuffs are not good? To the contrary, sugar provides us with energy that allows us to talk and work, while salt constitutes an element called iodine which is essential to our health and if the body is lacking it, one may have a health condition called goiter. In a couple of articles in this book, I have endeavoured to elaborate how fertilizer is taken up by the plant in the soil. I emphasized that for every amount of element taken up by the plant, it releases exudates in order to maintain the plant osmotic pressure. It is similar to humans; you can't keep eating if you are not relieving yourself in the washrooms or pit latrines. Immediately one stops visiting the pit latrine for a minimum period of two days, the next important building one will be trekking to will be the hospital (UTH) or the nearest clinic to seek medical help from the doctors. The most detriment fertilizer to the soil is nitrate fertilizers. This fertilizer has the potential to cause soil acidity if we are not carrying out remedial measures in its use.

Nitrate fertilizers consist of nitrogen which is a very important element for any plant to grow. Without nitrates in the soil for plant uptake, the plant will not grow; it will turn yellow and die back. Some people have turned to organic fertilizers like manure to supplement the nitrogen rquirements. If one is able to produce a lot of manure to put in the field, then well and good; they can take take that route, and there are some farms that are strictly practicing organic agriculture. However, we have to be alive to the fact that plants do not only need nitrogen for effective growth but there are all arrays of elements such as phosphorus, potassium, zinc, boron, iron, copper and many others which are needed for its effective growth. Some of these elements might be found in the organic fertilizers in the wrong amounts or they might just not be available at the right time when we need them. There is a process that takes place in decomposition of plant material to release these nutrients and it is called mineralization. There are some conditions that needs to be met for mineralization to take place. Therefore, to attain the productivity that we want to achieve, we cannot run away from inorganic fertilizers. The secret that I have always shared is that we need to routinely take samples from our fields for laboratory analysis to know what elements are lacking and what elements are in excess. The soil pH is a yearly requirement that we need to determine at our farms. The determining of the soil pH is like taking the body temperature of a sick. If we get a temperature of above or below 37^0C, then the doctor is able to infer or know what could be wrong with a patient. Once this is known, the next thing will be to rule out other possible diseases through a process of elimination by conducting further tests. It is similar with soil pH, if on the scale, you find that it is below 5 then you will know what elements found in the soil may not be available to the plant for uptake. Therefore, no matter what we say about fertilizers we have no option but to use them but the form could be different. The only thing we need to do is take remedial measures such as liming our soils if soil pH falls below 5.0 for instance. Do not be deceived; fertilizers will not destroy the soil if you are using them rightly. I have always given an example of York farm which does all year-round production and they use relatively good amounts of fertilizer but they have never abandoned any piece of land because it has been destroyed by fertilisers. For your own information, the average usage of fertilizer per hectare in Sub Saharan Africa is less than forty kilograms as compared to about 190kg in developed countries like Europe and America. Fertilizers are very important for crop production and in Zambia, there so many fertilizer companies that one just need to choose the one they want to use. I have heard that some fertilizer companies mooting to establish plants in multi agro-economic facility zones (MFEZ) that are being developed, and what are these multi agro-economic facility zones?

Multi agro-economic facility zones

The other time when I had a privilege to travel to Lusaka for work from Chipata, as usual the air in the city was full of a foul smell. Surprising enough, the corridors along Lumumba road opposite Auto World Spares had black pools of stagnant water and just next to those pools were '*sido*' (freelance mechanics) working on several motor vehicles (mostly suspension systems). The city is so congested that one would want to fly out immediately if going there for the first time. There have been a lot of shopping malls that have been built though; unfortunately, they are all concentrated in the busy central business district (CBD). I was perplexed to learn that the famous 'Tall Trees' which was a popular spot with the night life has been bought and another shopping mall is to being constructed there. Opposite it is the famous Makeni Mall at the junction of Kafue and Makeni roads. I thought to myself that the town planners of yesteryear are similar to our modern town planners. Why should they keep on creating more congestion in the CBD when we have plenty of land elsewhere? It is high time that they thought of twinning Lusaka with Kafue town or Chongwe to reduce pressure on Lusaka's Cairo road. The other day, the sewerage company was complaining that people should stop building septic tanks and connect to the sewerage system instead, but they are the ones that don't provide that service to the people proactively. Do they think people will wait for donor aid to build sewerage system to connect to when putting up housing structures? No, people have no houses and they are in a hurry to build. Underneath Lusaka is a very big 'sewerage system' and sooner than later, the whole groundwater may be polluted and it will be very costly to rectify it once it happens. The local council should be proactive; besides people pay for the services at the point of buying plots. Anyway, so much about the Lusaka and its reverse development, let's talk about real development that is being spearheaded by our newly found friends, the Chinese MFEZ.

In November of 2012, I read about this German company that is opening up an estate in Mumbwa (which am calling the Multi-Agro Facility Zone) and the vice president had been to the site to inspect the developments taking place. These are the agro-clusters that we need for economic development in our country. I was really happy when I heard about that development. For those that had been following me keenly with my articles I used to write for Times of Zambia, I had written about what once happened at Mpongwe Development Company, which we have ably discussed even in this book in the earlier chapters. Mpongwe district's economy is what it is because of the developments brought about by that company which was established as a scheme in the late 70s or early 80s (it should have been in 1978 or 1982). You will agree with me that my relatives, the Lamba's were not good at agriculture but because of what they learnt from that company, a lot of them are now successful farmers. in Machiya, Ndubeni, Malembeka, Lesa and many other chiefdoms have prospered because of agriculture. I liked the model Mpongwe followed of vertically integrating their activities and products. They once were the largest producer of maize, soybeans and coffee on the copper belt (even nationally). Mpongwe employed not less than twenty (20) first class university graduates and thousands of general workers. It was also used as a training ground for new graduates - I did my attachment there too. They were growing maize, soybeans and wheat, and milling it into mealie-meal and flower; by so doing they created so many more secondary jobs. If we had ten estates of the size of Mpongwe, we can even forget about drilling in the ground for minerals as if we are moles.

My hat goes off to some farmers in Mkushi who have followed a similar model, just like ZamBeef. We have also several companies that have come on board except that most of them are just crushing and exporting the raw material. Take for instance, the many factories that have been opened for extracting edible oils; many of them are crushing soybeans and extracting oil. The cake that remains is being exported as a by-product of oil extraction. I should commend companies like Tiger Feeds, National Milling and Novatek that are making cake into finished animal feed. A lot of companies know that they are not allowed to export soybean as a commodity, so they crush it and export it as cake. In my own opinion, we don't need such investments because we are just used as productive tools for raw material as the case is with copper.

This new company I'm told sits on a 10,000ha and from the little information I had gathered, it will be the biggest in the country. The company will not only be growing the various crops but it will be adding value and producing finished products. If what I have heard is something to go by, they will be bigger than ZamBeef. I am told that this company has also built a school; I guess they will build a hospital and possibly sports facilities. I will urge the government that they should encourage such investments and not what we had been seeing in the recent past where investors come to sell chickens and nshima at Soweto market and someone boasts of bringing in foreign direct investment (FDI). However, such investments come at a cost to the locals; the government should ensure that the locals that are displaced are well relocated in good agricultural productive lands. They should also ensure that manual jobs which will be created goes to the locals and Zambians should also occupy management positions like was the case at Mpongwe. I would be the first one to denounce the investment if I hear that accountants, farm managers and plumbers have been imported from German. The government should also take a deliberate policy not to allow them buy any more land than the 10,000ha they have. In case, they want to expand, they should be encouraged to find alternative land in Kaoma, Kabompo, Kawambwa or Kaputa so that development is not concentrated around Lusaka only. They can even go to Lundazi. We need such ventures more than we need opening up of mines and my hope is that the people are genuinely farming and not engaging in other activities which the government might not be aware of; like illegal mining of precious stones. Lastly, let me end by congratulating Hon. Simuusa for banning the indiscriminate cutting of timber – hardwoods (when he was the minister). I happen to have been to Kaoma, Lukulu and beyond. I was saddened at the level of deforestation which is taking place there. What annoyed me most was that some logs have been left to rot without being transported to Lusaka for processing. On average, it takes not less than 30 years for those trees to mature. The policy must be enforced that every logger should plant and nature some trees equivalent to what they cut. Just as I was concluding this article, I received a call from a Mumbi asking me to write about subsidies in crop marketing. I hope he will like my line of thinking because I write with an independent mind.

Subsidizing crop marketing

On Friday, the 26th October 2012, I was driving from Indaba village somewhere in Mwami area of Chipata where I had gone to attend a meeting by women groups on community banking famously called SILC. As usual, I decided to tune in to Parliament Radio to listen to our members of parliament (MP) debate various policy issues in parliament. It was exactly 12:45hrs and I suspect Hon MP from Eastern Province was the one debating. He started well by blaming the government's insistence on thinking agriculture is all about maize production and marketing; the sentiment I have always strongly been against. However, the member decided to divert and delve into some issues I was made to believe he did not understand well. He started talking about subsidizing marketing of other crops other than maize and not production.

I was uncomfortable when he talked about soybeans, groundnuts and sunflower. According to him; he said production of these crops is not a problem but what has been a problem is the marketing. I chose to vehemently disagree with the honorable member in that as much as he would want the government to help the farmers with marketing of these crops, productivity of the said crops has been a big issue. When we talk about the government helping the smallholder farmers with marketing, I don't think our members of parliament means the government subsidizing marketing like they are doing with maize; they distort the whole process. If indeed, that's what they mean, then they are very WRONG and they need to check their policies. Allow me to tackle this sensitive issue in this book from two angles; production and marketing.

Are we producing enough crops as smallholder farmers in Zambia? The answer to this simple question is no? Can the honorable member mention one crop which the smallholder farmers are producing efficiently apart from the horticultural crops; none! We have a lot of maize in this country because we have several hundreds of thousands of farmers that are growing it. The average yield of maize for the smallholder farmers is around 2.2Mt/ha against a potential of 8^+ Mt/ha. If a farmer is producing 2.2Mt and is buying all the inputs at full cost, that farmer will not break even if they sell their produce regardless of the market (FRA or Traders). To break even, the farmers need to produce at a yield not less than five tons per hectare (5Mt/ha).

Talk of sunflower; the demand for sunflower as an alternative input in livestock for feed making is enormous and if there is anyone who has 100Mt of that commodity, I can show him where to take it; he will be given cash at the spot. The average national yields for sunflower stands at 0.5Mt against the potential of 3Mt. Talk of soybeans, the story is the same. Smallholder farmers' average yields are between 0.6 – 0.9Mt/ha. This is far much less than what the commercial farmers are getting at 3.5Mt/ha and above. The story is worse with groundnuts because for this commodity, we do not even have certified seed in the whole country as earlier alluded to in this book. Whatever farmers will produce will be grown from either grain or quality declared seed (QDS). We need to support production as well. When we talk about supporting production, I don't mean to give subsidized inputs and handouts, NO! What I mean is enabling the production environment such as investing in research and development (R&D), effective public extension service delivery, promotion of better agronomic practices such as liming the soils and many other factors. It is quite disheartening to note that we are still relying on technologies that were developed fifty years ago such as the application of fertilizers using a blanket rate. Talk to any extension officer near you about production of maize. Ask them how much of the fertilizer one is supposed to apply in a hectare? The answer they will give you is four of the fifty kilograms D Compound and the same number (4) for Urea regardless of the type and status of the soil. Even in the alkaline soils like those in the valleys of some parts of Southern province, they will receive the same recommendations as the acidic soils of Mbala. No, that is not the agriculture we want in the 21st Century or fifty years after independence as the late president Sata would say.

Let's switch and talk about marketing which the honorable MP emphasized. I'm in total support that we should help our smallholder farmers to market their products, but what is important is the how are we going to do it. I know that the MP was thinking of fixing prices for groundnuts, soybeans and sunflower like they do with maize. That is not sustainable and this country doesn't have enough resources to compel the government to buy. What we need to do is to promote value addition through development of a strong agro-manufacturing base. If we promote localized processing industries in the areas of production, then we are stimulating marketing of these commodities. I should invite the MP to thoroughly study the soybean value chain and he may report to me what he will find. Soybeans have a ready market because we currently have about 300,000+ Mt of processing/crushing capacity of the grain. ZAMANITA which was later bought off by Cargill, Tiger Feeds, National Milling, Mt Meru, Emman, Global Industries Limited (GIL) formerly Gourock; name them are all factories that are using soybeans as a raw material.

That is what we need to do with cotton, sunflower and groundnuts; to name just but a few. In addition, am excited about the 8,000km link Zambia road project that the government had launched and my prayer is that it is done as soon as they can. When we open up all those 'dark' areas like the road to Vubwi, the road from Chienge to Kaputa, the road from Kaoma to Lukulu to Kabompo, the Lukanga swamps; then we are helping with marketing of the agriculture commodities. We can't depend on borrowed money to be pumping into unproductive activities (for consumption), NO. Zambia needs to improve the communication network as well in order to reduce the cost of sending a text message, increase power generation thereby reducing the cost of irrigating crops for instance. By so doing, we shall be helping the smallholder farmer to produce and market their commodities efficiently. Let the MP talk to the banks to reduce the cost of lending. If traders can access affordable finance, they will procure more crops and sell it to Congo, for instance. Let me end by stating that I was impressed with the debate by one Hon MP for Mafinga then (Hon Namugala). I wish she could have walked the talk when she was still holding power in government or was in the corridors of power because she debated to constructively without wearing an political jacket; it's never too late though. By constructively debating, you are contributing in a way to the development of this country although she could have done much more than when she had an opportunity.

One sector which is struggling is the cotton industry. This sector used to be the second biggest commercial crop after tobacco a few years ago, but now it is barely limping. Probably, it is prudent that we discuss this sector as well.

Agriculture without subsidies, is it viable?

The year 2013 marked the turning point for most of us in many things that we do in Zambia. We have to change how we manage our small budgets. Upon hearing that the government has removed subsidies on fuel, my friend was swapping his Toyota Ipsum 2.4litre car with a Toyota Vitz 1.0 liter just to manage his fuel budget. To be frank, I didn't know that the government was subsidizing the fuel in this country until it was announced. I think the main reason was that our fuel was and is still the most expensive in the region. I don't have better reasons why it is like this apart from insinuating that because we don't have a port. Nonetheless, we should also look at how we cost our fuel; maybe we might just find some unnecessary costs that make our fuel too expensive. By the way the cost of fuel does not only impact on those who drive but everyone including the unborn baby in the womb. It affects the common man moving on foot in Kaputa even that poacher in Kafue National Park. Well, let's not dwell so much on fuel subsidies because this is water under the bridge as far as this PF government is concerned. For now, let's discuss whether our hardworking farmers will manage their production without the fuel and maize 'consumption' subsidies.

It is a well-known fact that Zambia's staple food is maize although the northern parts of the country supplements this with cassava. The smallholder farmers produce over eighty percent of the maize grown in the country. We are literally fed by the smallholder farmers. Of the 2.5million tons of maize that was harvested in 2013, over 2.0million tones was produced by the smallholder farmers. The government supports the production of maize by subsidizing the maize inputs. However, this support has been mis-managed such that it has mainly benefited a few corrupt individuals. I would roughly estimate that sixty percent of the inputs lands in the unintended hands or wrong hands. When the late Mwanawasa introduced this program, the intension was to give to farmers for a period of three seasons and thereafter, the farmers needed to graduate and start buying their own inputs. This has not been the case; the same people have been getting the inputs and the unfortunate part is that many of them have been selling to commercial farmers. Of course, there has been a few that has put them to good use and have managed to increase their household assets from the sales realized such as buying implements, cattle, building good shelters and many other things. Now that this government has decided to do away with these subsidies, are the smallholder farmers going to manage?

My arm chair analysis is that there will be a slump in the production of maize at least for a season, which will definitely increase the cost of maize. Nonetheless, the double subsidizing of production and marketing of maize in this country was not sustainable. If I had a way, I could have loved to see the government continue for a few more seasons with subsidizing production especially that they have removed the subsidy on fuel. By making fuel costly, this will push up the cost of production and not only for maize but other crops as well. If not well managed, there might be a spike in inflation especially when the public service salary increment is implemented in September 2013. What we now need to do is to make a farmer an efficient producer of the crops that they are engaged in. We need to avail resources to the extension officers to ensure they train the farmers on modern methods of farming. There shouldn't be any farmer that should be producing less that 5mt/ha of maize, 2mt/ha of soybean, and at least a ton of groundnuts and sunflower. I am happy to share with the smallholder farmers that one farmer that is practicing conservation agriculture in Lundazi managed to harvest 16bags of soybean in a Lima (0.25ha). This translates to 3.2mt; a yield which is synonymous with commercial farmers that supplement with irrigation. This is evidence enough to show that even smallholder farmers with good management of the crop can get commercial yields. Secondly, it is time to do away with mono cropping – growing of maize on the same piece of land every year. Farmers should remember to practice crop rotation, which is, rotating leguminous crops such as beans, soybeans, groundnuts with cereals. The legumes fix 'free' atmospheric nitrogen from the air into plant absorbable nitrates, and this can reduce our cost of production for the subsequent crop. Farmers should learn not to burn the crop residue (Stover) after harvesting. The crop residue helps to build up soil organic matter and this is the source for some of nutrients for subsequent crops as discussed in earlier chapters of this book. Soil organic matter also helps to reduce the adverse effects of soil acidity and erosion. This can only be explained to our farmers if we have an effective extension service that has resources to visit each and every farmer in their farming camp.

Our farmers should also learn to diversify their production. Mind you, agriculture especially one which is dependent on rainfall is very risk although there is a limit to which we can mitigate such risk. In 2012, America did not have good harvests even if they are sophisticated in production because they generally received poor rains. Dams had dried and they could not use irrigation because the groundwater table sunk low. Additionally, we should run away from growing maize alone. By the way, these other crops like soybeans, groundnuts, sunflower and others are doing fine as far as having access to markets is concerned. Eastern province produces the most sunflower in this country but there is no commercial company which I know is buying this sunflower. What the farmers are doing is process this sunflower into cooking oil and sells the oil within their communities. The farmers are getting more money from selling cooking oil than they could have if they sold it as sunflower grain – this is what I call value addition.

My good guess is that the farmers will manage without subsidies but the first season will be very difficult although the cost of living will go up. This does not mean that we shall not have casualties; think of those that are barely making ends meet with FISP. What will become of them without it? The government should put an alternative plan for relief food especially next year. They should also show that they want to sacrifice with us by reducing its cabinet[10]. Honestly, I personally do not see the value some of them add to leadership and management of this country. I hope the resources freed from fuel and maize subsidy will be put to good use, although I have very big doubts because of the high level of corruption in this country. We might just end up benefiting a few that are close to the resources like what happened at the ministry of health some few years ago, where money meant to buy ARVs was misapplied. The other biggest mismatch I have observed in this country is the way we allocate land especially one that has been declared as settlement schemes. I can't understand as to how the government can de-gazette a piece of land as a resettlement area and target the retirees only. Do they know that agriculture is an activity which needs people

---

10 More ministries were introduced instead

with energy? You can't expect to have someone that has worked in other sectors and when they reach sixty years and have been discarded in their sectors where they were to go into agriculture and become productive. How do you only give productive land to retirees? Are you sure they are the right people for this energy demanding career of farming? I find this political decision we made concerning land distribution quite lopsided.

When the government finally made the decision to remove subsidies, a lot of 'dust' was raised in this country such that some people even resorted to take to the streets. I guess the reasons why there was too much talk, is that a lot of us don't understand well what it means. To a layman, removal of subsidy on fuel is increasing prices of fuel while it is interpreted as refusing to give farmers free inputs on maize or increasing the cost of mealie meal. It partly means just that because it is the government's role of ensuring that every Zambian has three square meals a day especially in a socio-capitalist economy like ours. However, it does not mean you just sit back and wait for the three meals to find you wherever you are seated. To be honest, I was very happy when the government decided to remove the K5 (it's actually more than K5 if we include transportation and handling cost, it could be around K15) per bag that the government was giving to the millers. Most of these millers are very rich now because they just sat back and waited for the government to do the 'donkey' work of aggregating (mopping) the commodity and go and buy it from them at a take away prices – just like that! We are talking of liberalized economy when we are heavily subsidizing the private sector –NO! It is a different story altogether if we turn to the input subsidy which the government has decided to maintain, however, farmers should know that even this will one day be removed; not too long away from now. What I will implore the government to do is to tell the farmers well in advance unlike doing things as though we never intended to do them (this has been the weakest point of the PF government as regards policy direction).

I should be quick to note that the input subsidy is not well implemented though. We have only looked at one factor of production – seed and fertilizer. To achieve meaningful productivity, it requires more than giving someone a bag of seed and two bags of fertilizer. The government should also look at how effective is the extension service delivery, what is the quality of the fertilizer that we are giving the farmers. I condemn in the strongest terms the way we are distributing the D Compound fertilizers which have a nutrient combination of 10:20:10 (NPK). For God's sake, these are not the only elements that are needed for a successful growth of a crop. What we need to do as extension agents is to be collecting soil samples hopefully every after five years for analysis and publishing these soil maps. The fertilizer companies then could be blending the fertilizers according to the soil tests results and the crops intended to be grown in those areas. I think Greenbelt and Zambian fertilizers have started this where they are making blended compound fertilizers. They are also adding Sulphur, magnesium, zinc and boron in addition to the traditional (macro nutrients) nitrogen, phosphorus, potassium and some organic matter in some fertilizers.

I guess that with the saved resources from the subsidy, the government will promote research for crops like cotton, sunflower, soybeans and groundnuts. For instance, we are seeing changes in the weather patterns where you sometimes have little rainfall and the other times too much of it. Our agriculture institutions like UNZA, Mt Makulu should be given resources to breed varieties of maize, for instance, that are drought tolerant that can withstand drier seasons and others that can withstand very wet seasons. We also need to equip our meteorological department with modern equipment so that they can give us proper weather forecasts (not just saying 'few....in areas, it doesn't mean anything). Are they able to detect a Tornado if it was to hit Zambia? (God forbid). Thirdly, the private sector has been crying that the government has been crowding them out when it comes to maize marketing because the government sets a very high price beyond the market price – they are right to a very large extent. We know the government has been trying to protect some inefficient farmers (those that produce less than 5Mt/ha of maize for instance).

What I would have loved to see would be the case were the government only buys maize at K65 per bag from those farmers that did not benefit from the input subsidy. This will encourage farmers to buy their own inputs so that they can access a better market, and for those that got inputs from the government, they should be asked to sell to the private sector who may be offering slightly lower prices than the government (K50). This is what is called smart subsidy and by so doing, we will be promoting a lot of farmers to be independent and others to produce efficiently. I know the question most of you will ask is that how will the government know that that particular farmer got inputs? It's simple, we have camp agricultural committees (CAC) and these maintains registers of who received subsidized inputs in each district. For those that did not get inputs and would want to benefit through selling to the government, the CAC would again maintain registers for such. For instance, for each one hectare of land cultivated, the government would only buy a maximum of 100 bags. Slowly, the government will move away from providing inputs. This FISP program has stagnated the development and investments of new fertilizer companies in the country. The traditional companies that have been supplying the government have been the only one benefiting since 2003. The system is prone to corruption and cartels in price setting.

We have been doing the same thing year-in year-out but there hasn't been any improvement in the quality of life of the smallholder farmer. Can't we see that the system is not working? Are we doing it because some of us (big fish) are the greatest beneficiary of this system? The argument of saying mealie meal will be expensive can easily be worked around by encouraging private sector investments into establishing of milling plants in areas were maize is produced. The government can deliberately support private sectors that are willing to establish milling plants in rural towns. For instance, here in Eastern province, we can have one miller based in Chipata to supply the whole of the province. Any mealie-meal coming from outside the province should be imposed with a heavy tariff to encourage retailers to buy from local millers. We are producing so much maize here in Lundazi and this maize is taken to Lusaka for milling and sent back to Lundazi as mealie- meal, isn't this 'madness'? Where is the business sense in this? Citizens Economic Empowerment Commission (CEEC) should be given deliberate instructions to support at least one milling venture in each province, and contracts with major retailers like Shoprite should signed with that particular miller in that catchment area.

Lastly, let me end this article by scratching on the fuel subsidy. It is a good move that the government has taken to remove this monster because the biggest beneficiary of such acts was the mining industry. However, it will be folly of me to think that just because I don't have a car, then I wasn't benefiting from this great subsidy, no I was! I would have loved the government to have phased out the removal of this subsidy by starting with big industries such as the mines and others. It could have been done over a period of a year before they completely remove it so that the impact would have been even. However, this will not mean anything if money saved from fuel will be used to offset recurrent expenditure like paying for the ever-increasing number of deputy ministers. The government should seriously seal all the leakages of such resources so that all the money saved should go to building infrastructure like the Katunda, Lambwechomba and Vubwi roads, developing newly created districts, schools and some social investments such as medicines in hospitals. I concur with prof. Saasa in one of the daily tabloid when he was quoted as having said that 'the success of removing the subsidy will depend on how the government uses this money'. Since we know how much we are saving, I would propose that the government makes investments into infrastructure of similar amounts per month. Sad enough, two years down the line after removing the subsidies, the government has borrowed heavily never seen before in the history of this country. We have not seen the benefit of removing subsidies; the leaders have been on free spending spree such that by the time we shall be having elections in 2016, we might just have over US$10billion in debts[11].

The cotton industry in Zambia

Cotton just like soybean, sunflower and groundnuts are cash crops in Zambia. The government has been subsidizing production and

---

11 Rumors has it that the total country debt had reached $17billion

marketing of maize for some time now. A lot of us have cried out loudly to the government that they should enable the business environment in regards to these cash crops. In Zambia, not less than fifteen percent of our national budget is financed by the multi-lateral and bi-lateral partners. There has been a lot of talk about subsidies in maize especially from the World Bank and other such institutions. We should not run away from the fact that even the developed countries like Britain, USA and others are subsidizing some of their industries including agriculture. However, they ensure that the subsidies work maximally to help the local farmers, industries and sector as a whole.

For a country like Zambia which has abundance natural resources but that are not well tapped to translate into economic benefits; meaning we do not have enough working capital, we should aim to offer smart subsidies to those sectors which will wholly benefit the people of Zambia – the productive sectors. Agriculture is definitely one of them but I don't share in subsidizing production of cotton. I am jittery about government subsidizing production of cotton because of many reasons. I would rather it lowers the production costs for cotton. When the government lowers the production costs for cotton, farmers will sell the commodity even at current lower prices profitably. This will mean the merchants in this industry will buy so much cotton and sell it at the world market competitively. Are we helping any Zambians with such policies? We are but not as efficient and effective as it would have if we had supported commodities like groundnuts and soybeans. The reason being that cotton is sold as a commodity; we might help to sustain a few jobs (about 200,000 smallholder farmers) but no local industries at all. The upstream industries that will benefit from our subsidies are the Chinese companies because most of our cotton finds itself into that great economy. I was thinking the government was going to consider putting those subsidies (production) into crops like soybeans and groundnuts because we have industries that process those commodities into finished products. By so doing, the government would have created more jobs than just wanting to sustain the same jobs. However, enabling environment must be created to encourage secondary industries in the cotton value chain. I am not saying they should not consider helping the cotton industries, but if I was in government, I was going to do it the different way. How? Cotton can easily be produced by small-scale farmers; actually, it is hundred percent produced by the smallholder farmers in Zambia. The best interventions the government should do are to create an enabling environment for the upstream industries such as Mulungushi, Kafue and Mukuba textiles to thrive. Let me guess that it is what our government will do if they really want to make agriculture the backbone of this economy.

By so doing, we can help create more jobs if our cotton is exported in form of replica jerseys for the Chipolopolo fans to Malawi or as designer's jackets to Bahamas and not the lint cotton, please we beg you (*twapapata*)! I have no doubt that this able government will rescind its decision if it was thinking of investing into production to sustain the foreign companies at the expense of our once vibrant Mulungushi textiles which is now being used as a piggery unit. They should think of operationalizing those cottage industries uncle Bob preaches on in parliament. You have really scored a plus with the Eurobond by investing most of that money into productive sectors such as power generation and distribution; that is the right way to do it because even if we don't immediately benefit from such investments, our children will come and praise us that they had parents that cared for them.

I urge our leaders not to envy some of their colleagues that just plundered and four years down the line, they start using the same plundered resources for court litigations, what kind of investment is that? We want farmers at Mpangwe to be adding value to cotton and sell it as fuzzy to some industry in Lusaka while they can also extract oil from the seed and sell the cake to the feed formulators. We can't continue selling a kilo of cotton lint plus oil and the cake at K1.60. *Awe, twakana*! (we have refused). Fifty years down the line, we want to still operate like we have just come from a war situation. This article was written in 2012, however at the time of publication of this book, the government had made some of the same mistakes I talked about in this article. They had promised that they were opening Mulungushi textiles in July 2016 but by August of 2017, nothing had been done. This is one biggest problem which has killed agriculture in Zambia; our leaders talk too much and do the opposite. Anyway, for now let me try and share with you my ideas of agriculture without subsidies.

Agricultural markets in Zambia

It is interesting to note how mis-information can be spread, and spread so fast. I couldn't believe my ears when I heard people were spreading false rumors after the removal of subsidies that subsidies are chemicals that are added to mealie-meal when being ground by the millers, and that those chemicals are very expensive and deadly in that they make people obese. I laughed my lungs out when I heard this. Some were insinuating that this piece of false information was being spread by those people who were supporting the removal of the much talked about and debated subsidy. Whatever the case, we had moved and life goes on with or without subsidies until the day we shall see our lord.

When this topic was still hot, I wrote an article in the one of the daily tabloids about the market opportunities for cassava. That particular week, my email box was full of emails seeking to know where the markets for various particular agricultural commodities are including cassava amongst many. It was evident enough that a lot of people were excited about that particular article on the prospects of a market for cassava. It was like opening a Pandora box because of the overwhelming response from the readers. It was really touching that the country did not have structured sustainable markets for many of the agricultural commodities that farmers produce. However, the truth is that anything that is grown in Zambia as long as it is food or agricultural commodity has a market somewhere (locally). It is just that the markets are not well structured and there is no systematic flow of market information in many instances.

Talk of groundnuts, maize, cotton, wheat, *Nyemba*, tomato, rice, cassava, soybeans, sugarcane, oranges, rape and *bondwe* (Amaranthus), name them; you will find a market. It is rare that there has been anything that has been grown and failed to sell except for some public institutions. For instance, in 2006 the government bought tons of cassava from farmers in Luapula province and all the commodity went to waste because they could not find the market for it. The latest had been the great harvest of 2010/11 which was called the bumper harvest of maize crop of over 3million tons. Thirty percent[12] of that crop went to waste (rot) in the midst of demand from the great lakes region and Zimbabwe. This is difficult to comprehend but sometimes the government just buy the crop to appease voters especially if it is during an election year, and indeed that was an election year.

The public extension staff, the NGO world and private sector have been preaching that farming is a business of which it is. In business, what do we do before we go into production? The first thing is to look for a market of any product or service that we want to offer, otherwise, if we don't do that, we will end up like the government who fails to sell the product that has been produced on a very dear cost. Before the richest man in Africa (Dangote) thought of establishing a cement factory in Zambia, he had to do a market research. He knew that Zambia's economy was a young one and it was booming, therefore, at the core of infrastructure development is cement. As farmers, we need to start to learn to plan our production and not to produce on impulse. Remember, in 2012 I was preaching that the prices for soybean will not fall any lower than K2.00 per kg regardless of how much will be produced. Most of the traders were buying this crop at prices between K2.00 to K2.60 farm gate prices countrywide save for those who do not conduct a market research. Mind you, we have had a bumper harvest of soybeans and sunflower that season especially in Eastern province. The previous year, we had a slump in the prices of cotton and a lot of farmers switched to soybeans and groundnuts but those that remained and grew it must be smiling all the way to the bank. In a very long time to come the prices of soybeans and groundnuts will remain stable because of the local demand for processing and the yawning regional markets such as Congo D.R., Zimbabwe and of late South Africa. The South Africans have opted to be importing soybeans and groundnuts than

---

12 ZNBC TV. (2012). Sunday Interview between G. Zulu and Hon. B. Sichinga.

producing it due to the high production costs as opposed to importing such commodities from South America and countries in the region including Zambia. After publishing this article, I got inquiries from countries as far as Asia who were looking for non GMO soybeans and I managed to provide that information to commodity traders.

As farmers, what we need to do is to plan our production by ensuring that we get the recommended yields for any crop. For instance, I have kept repeating myself that if we did all things right and got a yield of over 5Mt/ha for maize, any price above K50 per bag will be profitable for us even if we bought inputs at the commercial rate. We should also learn to diversify; do not just grow one crop. By diversifying, we shall be mitigating the risk factor as farming is a very risky business especially if you rely on rainfall. I have liked the initiative of most of the farmers from the east; they have a bumper harvest of sunflower (Zambian standards) in the province but none of the farmers is crying that there is no market. What most of them are doing is to process the commodity into cooking oil and sell within their communities. By so doing, they are making more money than they could have if they sold it as a commodity. Secondly, farmers should learn to be keeping records of their production so that before they sell their commodities, they can calculate the gross margins based on the prices the markets could be offering and determine what their break-even prices are. It is only by doing this that they will know whether they will make a profit or not.

However, they should bear in mind that to make a profit, each crop has a minimum level of productivity under which their yields should not fall below. This information can easily be accessed with the help of the extension officers in the area where you are. I am so impressed with the development that is taking place with regards to three crops: soybeans, sunflower and groundnuts not to forget maize because this is a *manchebele* (maize) country. We have learnt to add value as opposed to selling them as commodities. I get sick to hear that the government is selling maize to Zimbabwe; why can't we sell them mealie-meal? We should also work hard to develop the cotton value chain. If we keep selling it as lint, we shall always be at the mercy of the world prices. Remember countries such as USA, China, India, Pakistan and many others control the prices through the laws of supply and demand. Zambia's contribution of this commodity is insignificant although we import it back into our country as finished linen in form of cheap clothes from China.

To sum it up, let me assure our lovely farmers that there is a market for any agricultural crop that we can dare grow as long as we can plan our production and scan for the markets well in advance before we produce. We should also learn to use the modern farming techniques so that our yields can be reasonable and by so doing, our commodities will be competitive. Additionally, we should strive to add value to whatever we produce. Adding value does not necessarily mean a complete transformation of our products but even sorting and grading can make a difference in the price that our products can fetch. As long as we remain attaining yields below certain thresholds; we shall never win and we will keep complaining about the markets. Just imagine, I have seen women making a living by buying a chicken in some butchery at K30 and they chop it into smaller pieces which they sell at major markets after cooking it such as the junction of Great North and Tuta roads at Mukando or Kapiri Mponshi. They make not less than K50 per dressed bird. The government, I hope will play its role of making the environment enabling such as grading feeder roads and upgrading certain important roads like the Kaoma-Lukulu, Nchelenge-Chienge, Chipata-Vubwi and many others, not to forget the lowering on interest rates so that we can all access cheap finances like the Chinese do. Then, the rest will be history!

Smallholder farmers and crop marketing

Every time of the year when we start approaching April, farmers start to panic. They crisscross roads looking for prices to sell their crops at and to which trader. It is apparently clear that when it comes to maize marketing, the farmers have no choice than mostly to sell to government through the Food Reserve Agency (FRA). The reason being that FRA normally offers slightly the best price in the country; they could only be second to some Congolese traders in Katanga province. It is most gratifying for the farmers that manage to access subsidized inputs through the infamous FISP.

An average smallholder farmer that has been implementing good agronomic practices such as using certified seed, planting at the right time and rate, using the right amount of fertilizers, weeding in time; normally gets around five metric tons (100x50kg bags) of maize and more. If the farmer used inputs bought from commercial input suppliers (not FISP), they will incur production costs not more than K2, 500 per hectare, and if they sold this to FRA at K65 per bag they will have a total revenue of K6, 500 if their yield was around five metric tons. This is a good income even if they factored in other costs such as labour and transportation. It is even much better for those farmers that managed to get subsidized inputs from government as their profits are slightly better. If one was to do a simple calculation for the figures mentioned above, a farmer is expected to be self-sufficient in maize production in a period of two seasons of receiving support from the government. However, this has not been the case because of many other factors in this value chain.

Most of the farmers however, have been having good harvests of not only maize but other crops such as sunflower, groundnuts, sweet potatoes, soybeans and many other crops but they have been struggling to make ends meet. Every year when it's around this time, they complain about 'brief case' traders that invade the farming camps and blocks to buy these commodities. What the farmers fell to appreciate is the fact that these traders provide a very valuable service; a ready market. Had it not been for these traders, the production of groundnuts for instance, would have died a natural death like has been the case with sunflower in other regions of this country like central province. This is because their government, our government which is the biggest market for maize does not offer the market for these other commodities. It is not that I am blaming the government for not doing this because this is the role for the private sector, however, government is supposed to provide an enabling environment for the private sector to thrive. The fact that the government dubiously sold our only market for sunflower in 1992 (ROP), meant that they were not creating an enabling business environment for the sunflower farmers. I know we can keep on complaining and complaining until we complain no more! Then what should we do as smallholder farmers since ROP is history; at least for this country?

Remember that we, the smallholder farmers are in millions and we feed the nation. We managed to feed the region in the last three marketing seasons (2013 to 2016) but even after having a big market of over 200million people in the region, we still remain smallholder farmers. There is only one thing we fail to utilize when it comes to crop marketing. Most of smallholder farmers in Zambia, including other SME businesses are very poor at keeping records. Remember, did our enterprise budgets for the maize in the beginning of this chapter? Therefore, before we think of what price we are going to sell our soybeans, sunflower, groundnuts, sweet potatoes and many other crops this season, let's look back and tabulate all the costs we incurred in producing a kilo of each crop that we have. This is simple; just add all the costs of production (inputs, labour, etc.) that we had incurred to cultivate a Lima, hectare or acre of whatever we have grown. After we do that, we must divide the yield we have got from that unit into the costs incurred and what we get is the price at which if we sold our crops at, we shall neither be making a profit nor loss but breaking even or just getting back what we had spent. So what we need to do is add a percentage of profit that we anticipate. Don't worry if the traders that are currently buying are getting the commodity at lower than the price we have calculated. Keep your crop well for the next buyer who might be willing to buy your product at a better price.

However, if you decide to keep the crop till further notice, remember that you will be incurring storage costs and these will need to be added to already incurred costs to come up with the current total costs. If only we practice our crop marketing and farming in this manner, shall we be making any profits. Then only are we going to be farming as a business. However, caution should also be taken into account that you do not expect to produce five bags from one hectare and expect to recover all the costs incurred, because if that is the case, then know that you are producing inefficiently and already you have made a loss. As we sell our crops any time we produce, let's be cautious and practice real farming as family businesses. Mind you, agriculture has the potential to contribute more to gross domestic product than it is currently doing, even more than mining because we as Zambians don't own any mine but we have thousands of farms or land on which we practice different type of farming. We need to be an example of the green revolution in the region and we shall keep reminding the government of their responsibilities. The responsibility to create enabling environments for all our businesses to thrive including agriculture.

Smallholder maize pricing nightmare

Firstly, may I take this opportunity to earnestly thank the hardworking farmers of Zambia for a relatively good crop that we have produced against all odds in 2015 cropping season. Crop production in that year was against the farmer because of various negative factors that were at the farmer's exposure. We had the Kwacha weakening against the major currencies such as the dollar which made the cost of inputs become so dear beyond the reach of many of them. It is not a secret that we import almost all inputs for production except for seed and labour. I was almost shedding tears when I was buying fertilizer for my small '*shamba*' at K380 per 50kg bag. The cost of herbicides for weed control; fungicides to protect my crop against diseases and insecticides almost shot to double the price as compared to the previous season. My cost of production was around K5, 200 per hectare for maize. Knowing that the season may have challenges with rains, I doubled my fungicide application just to give my crop that added protection in case of eventualities. This worked to my advantage as I managed to harvest 6.8mt per hectare; this was not a mean achievement considering that I depended on erratic rains that season.

I know that for most of the average farmers, especially those that apply fertilizer as though they are putting salt in Kapenta (small fish); they averaged 2mt/ha and less. Farmers are in business of making people develop pot bellies and if they don't operate as a going concern, people might start looking as though they live in a war torn country. Regardless of the production environment, farmers have to recoup the costs and make at least some profit so that they don't fail to buy inputs for the following season. During the marketing season, I normally get surprised because farmers start to ask for the price they need to sell their crops at! They are the ones that produced and are expect to know what price they need to sell their products at. The government had announced that they would buy the maize for strategic reserves at K75 per bag. For a person like me who got over 7mt per hectare, that was not a bad price because I ended up making a gross profit of K5, 270. However, a less efficient farmer like my cousins made losses of about K1, 500 at that price. This meant some ended up not participating in the subsequent season because they failed to buy all the inputs. This does not mean they should sit back and cry that *Boma iyanganepo*, (let the government intervene) no! Governments normally don't do business; they needed to search around for the private sector that I am very confident that they were offering better prices than that. I expected them to have bought maize at between K80 to K110 per bag. This is not a political secret because we were the only country in the region with surplus corn and all countries had their eyes almost popping out of the sockets; they all focused on Zambia to be their savior. We will never be any luckier than we were that season because the warehouse receipt system (WRS) was then in effect; if one does not agree on the price offered by a particular buyer, they just needed to deposit their grain with any certified warehouse and wait for the price which might satisfy their level of productivity – that is business.

As for the government, I do not blame them on the prices they set for their purchases because if they had upped the purchase price, it was going to also have a ripple effect on the price of mealie meal. They act as a social corporate responsibility for most of the poor that might not afford the price of food. Therefore, commodity pricing in that particular year had been made very simple because of the demand for the commodity.

Some people had suggested that the government should have maintained the normal 500'000mt strategic reserve as opposed to the

1'000'000mt[13] which they had indicated to buy. I had a different view to that. That one million was enough for two years with the current consumption levels. I was of the view that they could have even bought for three seasons (1.5million MT). I guess that is what we were all talking about in January when we had that shortage of food. Suppose we did not receive good rains of that season, where were we going to get our food? The justification that it might be subjected to misuse by the people tasked to look after the reserves did not hold water for me. I think Dr Kayula and I need to sit down so that we come up with strategies that will ensure that our reserves are not subjected to misuse by people we have employed. Mind you, agriculture in Zambia is dependent on rainfall, which we have no control over. In most cases, I have observed that what farmers lack in commodity pricing is information besides poor record keeping. This information also comes with the level of education, however, there are some farmers that are well educated but behave as though they are not. Is it that they are schooled and not educated? What role does information play in agriculture?

The 'power' of information

In the last five months of 2015, my inbox had been inundated with people looking for information about the agribusiness environment in Zambia. I received requests from as far as South Africa, France, Zimbabwe, Turkey, Botswana, and India and even within Zambia. What surprised me was that some of the information they were looking for were too basic; information that is supposed to be in public domain. When I asked those from Zambia that they could access such information from ministry of agriculture and Zambia Development Agency (ZDA); they responded that they had been to those organisations but the information provided was no so helpful, and I wondered but why!

To attract investments in agribusiness we need to package information which has to be up to date and relevant to the needs of

---

13 The government failed to buy even the normal 500,000Mt that they buy, they only managed 230,000Mt

would be investors. For instance, if one had to go to some of these public offices mandated to formulate policies in agriculture and ask for the production costs for cabbage, I am so doubtful whether they will have the accurate information. The only problem we have in this country is that we think agriculture is all about maize production. I am challenging NAIS or is it the farm management department of the ministry to package such information and update it every so often. I really felt bad when this person told me that the information provided to him about the production costs did not make sense as it was 'ancient'. When I analyzed the information, I realized that they had supplied him with smallholder costs for maize production (to say the least it was outdated) even after telling them that he was looking to acquire 5,000 hectares to go into maize, wheat, soybean and potato production. They need to supply him with commercial costs of production which was going to help him with decision making. You might argue that information is costly so this person needed to buy it; yes, he should have parted away with a few dollars if he had gone to a private company but not the ministry. We pay the ministry to be providing such information to people like him and anyone else that want to venture into agriculture. Besides, how was he going to buy information that was not going to be useful to him? As an individual, I supplied them with information that they appreciated so much, what more the ministry? It should have been much better and accurate.

I think going ahead, the ministry especially the department of information (NAIS) need to restructure their operations and start looking at agriculture with 'new' lenses. They can task two to three people within their department to be regularly collecting this information from other stakeholders. For instance, if today I was to ask the ministry what the national consumption for herbicides or fertilizers is, they will be giving me ranges for fertilizers and they will be blank on herbicides. This is information that they can easily get, how? They can go to ZRA and ZEMA and collect information on all companies that bring in herbicides and fertilizers for instance. Thereafter, they can go to individual companies and get this information, which is consumption because not everything ordered, is used that year. It will be very difficult for an individual like me to get it from such companies because they might think I want to get trade secrets and use it against them to gain competitive edge but for the ministry, they know that it is their mandate to provide such information as long as they don't disclose such to competing companies. I remember in 2012 while working on a project in Eastern province, we were looking for costs of production for onion and tomato. We never found that information and we were forced to start calculation using primary data but that is information lying with companies such as Amiran, ZamSeed and others that are involved in vegetable production including farms. The department should foster strong partnerships with private companies such as Amiran, Bayer, Syngenta, Omnia, Yara, BASF, CHC, NWK, AFGRI, ZamSeed, Monsanto, SeedCo, banks and so on and gain their confidence so that they can regularly be furnished with such useful information on a regular basis. Honestly, it would be very difficult for someone who has just come from Turkey to get production costs for particular crops if they have to individually visit these companies. It would be easy for them to access that information from the ministry and quasi government institutions such as ZDA. I guess we have concentrated so much on maize such that we have ended up ignoring other important roles that we need to be providing as the ministry.

To help my brothers for information that I could not provide them with, I ended up directing them to IAPRI and some private companies but I felt bad because this was information that was supposed to be at fingertips of NAIS. Information in business is very important and it is the only panacea for decision making. Let me urge the ministry responsible for selling land in this country that they should also be careful with the way they are disposing off our land. Our brothers from South Africa and Zimbabwe are eyeing Zambia for agro-investment. They are most welcome but we should not give out larger junks of good land at the expense of the locals. My advice would be that they need to put up legislation to limit how much land a foreigner can own (not more than 1,000ha). Anything above that has to be done in partnerships with locals. I guess that with good information, farmers will be guided on what to produce and which market to target.

Markets must guide farmers

The time between April and August of each year, the farmers' incomes increase exponentially. Needless to say, real farmers get money every day as they spend it; meaning their farming is throughout the year as they know that farming is a business and a career just like the bankers, engineers but of course not politicians. Many of us are already knows what crops we will plant by May of each year. A thrift businessperson in agriculture will think ahead knowing of what to produce for the following year based on the scanned market information and trends. For instance, in an election year, politicians normally become so generous that they don't follow the laws of demand and supply. In 2015/16 marketing season the price of maize was pegged at K85 per 50kg bag while in the following marketing season, it dropped to K60 although the cost of inputs kept going up. So, as a farmer, what would you produce for the following season?

This is a simple but difficult question at the same time in that you don't know what will be on high demand next year, but normally, it will be prudent to spread your risk. Do not put your eggs in one basket; I mean don't just plant one crop just because you anticipate a good price for that particular commodity the following year. Having analyzed commodity marketing especially in Zambia for some time, I can roughly predict that the price of tobacco, for instance, might surge up in next year (2017). The reasons being that not so many farmers had planted tobacco in 2016 because of the currently low commodity prices on the market. Politicians normally are very good to us when they are asking for a vote and they have come to know that we love our nshima so much that if one has not had his or her 'five fingers' regardless of whether one had other foodstuffs means they have slept on an empty stomach. The price for maize will be relatively good and inputs may be easy to access next year (2016). This marketing season, we have seen excellent soya prices and this could be attributed to low yields generally experienced by farmers whilst the crushing capacity has tremendously increased in the country. Farmers that are not growing soybeans should think twice as this is the only alternative they have to practice crop rotation besides the traditional groundnuts. I have always preached that soybeans are one of the few commodities that has vertically been integrated in that we are able to not only extract oil but can make animal feed, and other foodstuffs such as the *nyamasoya*, soya chunks, milk and many other products. This is good for our industry unlike other commodities like cotton where we export it in its raw form as lint.

When you plan to grow any crop; do it like a professional as we can't afford to waste any resources including water? As a farmer, we can't afford to replant as by so doing, you will be losing some yield. You may ask to say as to how you can't do it when you depend on rainfall which we have no control of when it will rain. Thrift farmers know that in times of erratic rainfall, it always pays to practice conservation agriculture such as minimum tillage, leaving crop residue as mulch, ensuring there are no weeds competing for the scarce water and nutrients. You can only do this by ensuring the use of herbicides or weed killers as some people would like to call them. There are so many weed killers on the market today but it would be good for you to seek professional advice of the type to use and when? A lot of farmers have fallen prey to 'crooks' that have either sold them expired or wrong products. What you ought to know is that weed killers do not control any type of weeds; they are weed specific serve for two or three that are not selective. Therefore, you need to know the weed spectrum in your fields. Some of you just know the traditional Acetochlor and Atrazine. However, there are other with very good combinations of actives that have a broader weed spectrum to control on the market. Nonetheless, you need to seek expert advice of what you need to go for in either case. Some of them also depend on the soil type you are farming on. Many of you have managed to control weeds so well but have ended up losing your yields due to your crop being attacked by pests such as worms/caterpillars or fungal diseases. In Zambia today there is no pest or disease that cannot be controlled or managed, including viral diseases which have no cure. You might be thinking am crazy; no! Viral diseases are normally transmitted by vectors such as aphids and if you can control the population of aphids then you can control this disease. Do not allow the disease to start showing symptoms before you apply your fungicides. Normally, in a season that we have doubts about rainfall, it is always good to apply *strobys* which helps the crop to manage stress. These again are crop specific; you don't have to apply anyhow. I normally call farmers as 'Pediatrics doctors' for crops; they deal with 'patients' who don't say what is troubling them. Before you buy your famous beer from the sweat of your harvest, ensure you have all that is needed for your next planting season.

The well organized dairy sector

In Zambia, there are more tomato growers than those that are involved in rearing dairy animals. It is estimated that over half of the vegetables including tomato that are sold in the chain stores in the country are mostly imported from South Africa. The trade imbalance in food items besides other products can easily be deduced by the number of trucks coming in loaded with products than those that are going the opposite direction. In most cases, the trucks going the opposite direction will either be empty or carrying copper ore or concentrates. In early 2017, this compelled the minister of agriculture to ban the importation of vegetables because the country was producing enough vegetables. However, one of the prominent vegetable merchant was the first one to cry loudest, claiming that the Zambian farmers were not producing enough and the quality was not to their standards. This was a raw deal the hard working farmers were getting. On a clip the national television station, the vice president of ZNFU was seen supporting the action taken by the minister. It was disappointing though that a few days later, the action was reversed after the counterpart at commerce complained that the country will be losing incomes in form of taxes. How much that was left to the economist to decide but even a lay man knows that the country could benefit more from producing than importing.

It is very difficult for the minister to convince any Zambian that the country does not grow enough quality vegetables that can be sold in these chain stores. In any case what has changed now, when since independence the farmers have been producing enough which the country has been fed on including tomatoes. The trend that has recently been developed with the help of politicians through policies that are not supporting local production in certain will one-day backfire in our own faces because certain countries where these foods are imported from embrace genetically modified organisms (GMO). The country's regulatory institutes may not have the capacity to detect such in processed imported foods. We have no doubt whatsoever to believe that even raw food can easily see itself through the borders of the country undetected. With this scary scenario, it is imperative that the population relies on its home-grown foods. That is the only way it will maintain its well branded of being a non-GMO growing country. The agricultural sector is therefore implored to be well organized. There should be great support, of course with the right policies from government. This sector may benefit by learning from how the dairy sector that is well organized.

In Zambia, there are more beef cattle than dairy animals but the country is self-sufficient in milk supply though there could be instances when some milk could have been imported from Europe. That could be taken as exceptions. The reason this has worked so well is due to the great involvement of a private company that has been the driver of the milk sector. Credit should all given to Parmalaat, a company originally from Italy. It is not known whether this company owns even a single ranch of dairy animals in the country but they have adequately supplied all the supermarkets and open markets with fresh milk which is bought from ALL KINDS of farmers. Emphasis is made on the type of farmers the milk is sourced from because this is the point that is normally being advanced by certain chain store that are always complaining that the quality of tomatoes, onions including cabbages that some local farmers have attempted to supply them is of quality. This is this quite odd because it is these same farmers that have met the quality and standards needed by Parmalaat in the milk value chain.

Taking keen interest in this value chain to study how this company has been able to meet the targets in milk procurements, it was so satisfying to learn that the company has invested and invested well in this sector. Fresh milk has a very limited shelf life than vegetables including tomatoes for instance. Therefore, the vegetable traders cannot complain about the quality being compromised when delivered. Interacted with these farmers, it was learnt that Parmalaat has really invested massively in the value chain in order for them to receive quality produce. The farmers have been trained on how they need to harness their product so that it can meet the standards and quality needed. They have organized themselves into milk cooperatives with help of other organizations as well; organizations such as SNV, Heifer Zambia and those projects funded by USAID. These farmers have even learnt complex operations in livestock management such as artificial insemination for breeding stock. They have been trained in feed formulation that can enhance cow milk production. They also have been trained in good milking operations and how to detect diseases or disease-causing pathogens in their product. Many of these farmers used to be large maize growers just a couple of seasons ago but they realized that with the ever-increasing cost of getting maize inputs and the changing weather patterns, they will be more competitive in milk value chain and switched to dairy farming. The ever-fluctuating commodity prices also forced them out of production of maize at commercial scale; now they only grow maize and soybean for feed formulation and food for their families.

On an agreed time in a week, Parmalaat sends a refrigerated truck at milk collection centres to collect the milk from the farmers. Some farmers boasted that they make as much as K30,000 ($3,200) per week from their milk supplies. This is equivalent to the money they used to make in a year when they were growing maize. This is one sector which is working so well and this needs to replicated in the vegetable sector and ban all the imports from South Africa and Tanzania. Other countries have done this to improve certain sectors that have been lagging behind. In this sector, there are no subsidies but it is working so well.

Input subsidies, have they delivered?

The period before 1992, agriculture in Zambia was taking shape. During that time, we had companies like NAMBoard and Lima Bank which were responsible for providing support to farmers especially the smallholders. The former was responsible for providing the market to maize and sunflower farmers, while the latter was giving financial support or loans to farmers. I should state that initially, the repayment rate was not as bad as it became in the latter parts of their existence of the two government institutions. Many factors could be attributed to this and some of which are due to nepotism as well as politicizing the institutions making people think public resources are not supposed to be paid back. This is one piece of culture that has killed public institutions in Zambia as well as Africa as a whole.

In my view, I think providing support to the farmers in form of loans is the most sustainable and effective way as long as we create an enabling environment for marketing activities of the churned-out output. My good guess is that the repayment rate became very bad during the period 1985 to 1992 when the economy of Zambia had collapsed. Later from 1992 to around 1994 when Lima Bank was liquidated and NAMBoard was disbanded, we saw the proliferation of other institutions like CUSA which were given the mandate to manage loans on behalf of government while ZCF and others were agents that were involved in crop marketing and handling loans on behalf of CUSA. You will agree with me that by then corruption was so entrenched in many Zambians such that government lost colossal sums of money through failure of farmers to repay loans. I remember agents resorting to last options of collecting assets such as beds, mattresses, ploughs and anything they could lay their hands on from defaulting farmers. Some of which were not even of value.

In 2003, the MMD government through the Mwanawasa administration introduced a program they were calling farmer support program (FSP) which later became the current FISP. The whole idea was that this program was to render support to vulnerable but yet viable farmers for three consecutive seasons with inputs such as seed and fertilizer. After the third season, it was assumed that the farmer would have been adequately empowered to enable them have capacity to procure their own inputs. This is fourteenth year down the lane; the government is still providing the support to the same farmers except that we have scaled down from a one hectare to half a hectare packs. The question many of us is what could have gone wrong?

I still feel that there are certain things that we did not do well and we have continued to do those things wrongly to date unfortunately. Firstly, I have a strong feeling that no matter how efficient a farmer is, cultivating a hectare to produce food to feed an average family of six members for one year and have surplus to sell in order to buy eight bags of fertilizer, a 25kg seed, herbicides and lime to apply in the one-hectare field is a gamble. My feeling is that with the support provided, it is not possible to consistently increase productivity to levels that can guarantee making a profit because crop production is not all about fertilizer and seed, there are other things involved such as herbicides, pest control chemicals, lime and extension delivery. Secondly, I feel that we have been targeting the wrong people for some time. For such interventions, we need probably to look at farmers that are cultivating more than five hectares if maize is the main crop we are targeting. We could have divided this program into two categories; FISP targeting viable farmers that are cultivating more than five hectares to receive support for three seasons and then for food security, this could have been channeled through the ministry of community development and PAM for those vulnerable farmers that we want to help improve household maize security. The latter are the ones that could have been provided with hectare pack as it were or half hectare pack. I have a feeling and this is true that this innovation has been marred with corruption. A lot of inputs meant for the farmers have ended up being sold on the open market by the people managing the program. Lastly, giving a Chinese a fishing line without training one of how to fish does not mean you have empowered one to be fishing. The provision of such support needed to have come with effective extension service delivery. I know the government cannot do everything by themselves but this is a government program and it needs to look at all actors and activities involved. The public extension service delivery is not that effective in that they lack resources. How about partnering with private sector such as channeling this support to be managed by a private entity like CFU, seed and chemical companies?

In a nutshell, I guess the program has not or partially achieved its objectives. We are only maize secure though at a huge cost but has failed to be on a revolving basis where farmers needed to be weaned off after three seasons. Had we managed to be weaning off farmers after three seasons, we could have been talking of the supporting getting to all farmers by now. The point I will emphasize again is that we targeted wrongly and we had no monitoring component of the program which is most the case with government programs. However, It's not a secret that governments world over offer subsidies to their citizens in many forms. Some of these subsidies are offered at production as well as the market ends of respective value chains. We have heard of certain interventions made in aiding the commodity prices remain competitive in developed countries by governments buying off excess production from farmers and donating such to third world or hunger stricken countries. In certain instances, some farmers are paid for not producing. This is the case with Zambia where the government has been subsidizing smallholder production by offering inputs at almost a third of the input costs.

In our country, we have seed materials that have the potential to produce over 10tons per hectare. We have seed companies such as SeedCo, ZamSeed, MRI, Pioneer, Pannar, Monsanto and many others that have good seed material. You will agree with me that most of these companies if not all of them supply these same seed materials to farmers through the FISP but the average national yield for small scale farmers is around 2.3t/ha, why is it like this? You may fail to comprehend that the same fertilizer and seed material when used by commercial farmers, they get yields as high as 12t/ha on the same land. What is the magic that the commercial farmers are applying in their fields which the small-scale farmers are not?

It is surprising and annoying to learn that we know the reasons behind the low yields but we have chosen to pay a blind eye. Firstly, we will never ever attain meaningful yields if we continue to be waiting to weed our fields only when we have seen the weeds in the field. For example, in December 2015 I lost two calves because my handler in his own right decided to be dipping the older animals and not the calves. In his understanding, the dip was going to be poisonous to the small animals unlike the old ones. If he was a worker, he could be on the streets now but he was lucky because he is a relative though he was strongly warned. I had always been buying the dip and I had been telling him to be dipping the animals and not the old animals only. Anyway, the point I am trying to drive home is that it is always good and cheap to prevent something than to cure it. Why have we been providing the seed and fertilizers only when we know that the farmers also need herbicides for them to get meaningful yields?

Our farmers have never graduated even when we have been giving them enough to produce one hectare. Adoption of using seed and fertilizer only is not good enough to increase productivity for our small-scale farmers. We as a people need to invest a bit more than just seed and fertilizer in our production regimes. The government should consider including herbicides as well as proper extension service as a full package to improve productivity in this country and many countries in Africa too. Additionally, farmers need to be made to understand that using seed and fertilizer alone will not make them achieve the 10Mt of maize that they desire to produce. Thinking from the top of my head, I think the government could have reduced the package from one hectare to half a hectare packs because it wasn't seeing any results in productivity improvements because the farmers were not managing to weed their fields in time, hence, they opted to only plant half of the fields and sell the other half in order for them to raise some money to use for labour in weeding their fields. Suppose, we had included a 1ha herbicide package, don't you think the farmer would have managed to weed their one hectare in a day? With a knapsack sprayer, I manages to weed two hectares in four hours (of course including the time I was fetching water). We have very good herbicides on the market such as stellar star and Lumax that controls both grasses and broadleaf weeds pre- emergent. I believe this program can do better with a bit of tinkering on how it is managed.

Creating agribusiness giants in the sector

Have you ever imagined a lion swallowing a tiger or an elephant swallowing a rhino and on the other hand a zebra 'marrying' a donkey? This sounds more fictious than possible but it is happening in the agribusiness industry. From the time, we heard of some agribusiness giant being taken over by a chemical company from Asia, a lot of things are happening in the background. Many multinational companies are looking out for whom next to buy or partner with in the agribusiness industry. I have been wondering as to why the drive to take over rival companies. If not greedy, my best guess is that the burst in the economy of China which ultimately led to the slump in many commodity prices had partially played a major role leading to marriages of conveniences. The global economy created a 'monster' in the Chinese economy and once things started falling apart, everyone else had been affected. We all had put our ostrich eggs in one huge basket and we are paying for our mistakes. There are so many countries with cheap labour forces in the world but we all rushed to China, though many of the companies complimented with the possible available market.

I was of the view that the China was a good experience but we are still going ahead to create giant input supply companies. I know that sometimes it is irresistible if someone dangles a juicy carrot but we should always think about what we shall eat tomorrow when the carrot is finished. I honestly can't imagine what a massive company will emerge from the successful takeover of the largest seed company by another largest agro/pharmaceutical company in the world. We are talking of an equivalent of 'USA merging with Germany' to become one super power, what a great super power that would be! I have no ill feelings about this anticipated takeover of the decade if I may call it but it will have negative ramifications on weaker 'souls' like the smallholder farmers of Africa. A couple of weeks before the takeover, I was so elated to write about the many multinational companies that were opening or establishing bases in Zambia. My thinking was that competition was going to be stiff and it was going to work out to the advantage of the less efficient producers who are in the majority. My celebrations were short lived. Think of it in this way, suppose PF and UPND merged today to be one political party, which individual opposition party do you think will dislodge them from power? Do you think ZDDM or MMD can be a match? Not at all! I felt deflated and my energy levels are slowly dwindling. I was of the view that the farmers of Brazil, Argentina and USA would look through that merger and object for outright take over. There are so many reasons for my apprehension besides making the cost of technologies unaffordable to Africans. The other reason being that these companies will become less innovative. Look at this example, one company came up with an herbicide suitable for smallholder corn growers called Stellar star. It is a very good product which is a great solution for weed control. However, another innovative solution provider responded with Lumax which is equally a very good herbicide for smallholder maize farmers. I am sure the other players must be in their laboratories to try and come up with something more innovative than these two products. This is what competition breeds – innovations! Just rewind the time some ten years ago, we had so many phones on the market such as Nokia, Ericsson, Sony-Ericson, and many others. People had a wider choice and they were more affordable but today if one thinks of buying a phone. It has to be Samsung, iPhone or Huawei; the other brands have been obliterated and it all started with one brand buying off the other. What this has bred is that if you want to get a good phone, one has to have not less than $500 when we used to get phones of similar quality at $100. Competition is being stifled and collusion is what is being bred.

Secondly, I have 101 doubts whether the giant companies being created will be any efficient. In their current forms, some of these companies that are thinking of merging or buying off others are already giant companies with so many strategic business units. When you think of them adding new production lines, I guess if they will be more efficient than they are. Just watch this space; twenty years from this date the amalgamating companies will be talking about disbanding or selling off other production lines. I have no problems with farmers from the developed countries but my worry is the resource poor farmers like those of third world countries of Africa. Some of these technologies will become so dear that even the new production frontiers most of them are talking about will be a pipe dream. Remember that Africa and some Asian countries are the only continents that are remaining with enough potential to feed the growing global population. We just need to keep working hard because the current commodity slump caused by our actions in certain economies will be over. We are just creating the 'witch weed' in agribusiness development of the world. I personally, I am against this drive but what influence do l have anyway?

Market opportunities for cassava

Did you know that Zambians, including myself like *nshima* too much? We get over ninety percent of our energy from *nshima*. Way back in time, I remember my grandmother telling me that as long as there is *nshima* one can't starve. As a matter of fact, she could prepare *nshima* and make salt solution for us to eat when there was no relish. We get this most treasured food item from milling most of the cereals and starch tubers such as cassava. I will not be surprised to see a Zambian grinding rice to make rice meal for *nshima*. However, do you know how many products can be derived from cassava (*Esculenta manihot*: Chinangwa, Tute, Kalundwe)? There are so many but here in Zambia, we only mill it and eat it as a snack (ubwaushi). In other countries like Thailand, they use cassava to make glue, paper, and it is also used as a filler in the pharmaceutical sector, and many other products. Just a stone throw away from where I'm writing this article from (Chanida border: Katete-Mozambique), SABMiller, the mother brewing company that owns the majority stake in our own Zambian Breweries is making beer from cassava. Why should I bother to write about beer especially deriving it from an important food crop like cassava? Do babies drink beer? These are some of the questions that many of you might be asking at this point as you are reading this book.

A colleague and I buying cassava at Nkenya market in Congo DR from a Zambian trader

Let's 'walk' the talk. In 2009, while working for the project in Luapula province that was trying to commercialize agriculture on selected value chains, I had an opportunity to facilitate some market linkages. One of the value chains we were working on is cassava. Luapula and Northern (then) provinces used to produce the most cassava in this country, followed by Northwestern, Western and Northern parts of Eastern provinces. We had challenges to find a sustainable market for cassava apart from the Congolese traders that would puddle across Lake Mweru and Luapula River to buy from some farmers in Luapula province. A few traders would take the commodity to Kitwe's Chisokone market where buyers from Congo and Angola would come to trade. This trade was very difficult to quantify because of the wide border line that Zambia shares with these two countries. In 2008 we had undertaken some market survey into DR Congo's open markets. It was gratifying to note that there were a few Zambian traders selling cassava in Nkenya and Zambia-Zaire markets. I should state that this was not significant. Most Zambians feared to take their commodities to Congo D.R. because of the fear of being robbed and too much corruption on the roads. It's not long ago when one of our own drivers was killed in that country (2013) – it's not safe especially if you are driving a Zambian registered vehicle (you might end up losing both the vehicle and your merchandise). What most of the traders do is to trade at Kasumbalesa border even though they know that the market across is yawning for anything including red ants.

As a project, we facilitated a market survey into Angola as well. That is another very big market for cassava. The findings; so interesting where that there is a very good market for cassava but the challenges outweighed the benefits. Some of them are that there is literally no road infrastructure from Zambia into Angola. It took our trader over two weeks to move thirty tones of cassava from Mwinilunga into a border town of Angola. Secondly, there was a very big barrier in communication; people need to start learn Portuguese and some Angolan languages especially now that the Kalabo road connection to Mongu is complete. The price in Angola at that time was four times the price at Chisokone market in Kitwe. These are very two big market opportunities Zambian Entrepreneurs in cassava have failed to explore and the biggest barrier has been the infrastructure. I will not dwell on these two markets in this article, but I thought of sharing with you my most valued readers about the potential that is yawning out there.

Former governor of Katanga Mr. Katumbi admiring agricultural products from Luapula at Katanga's trade exhibition

As a person tasked to facilitate market access for the people of Luapula in that project, I had to look for alternative markets with quick results (funded development projects are short term). I contacted someone I will not name from SABMiller South Africa whether they could buy cassava to be used in their brewing industry in Zambia. This person was keen to know the production figures (excess of home consumption) that are produced in Zambia. It is very unfortunate that this country does not take data collection and management very serious. In business, if you report to someone who wants to buy a particular commodity as, "production of cassava in such an area is about...", they will not take you very serious. When I went to the responsible institutions that are tasked to collect this data, I couldn't get meaningful data to build my case to convince them that we produce enough cassava in this country. As a matter of fact, I can challenge anyone to give me the value of trade between Luapula and D.R. Congo; no one will provide me with that information. Unfortunately, I started pursuing this activity towards the end of my contract and I wouldn't know how much this agenda was pushed by my colleagues that remained behind. What I want to write about in this article is that I am very happy to have listened to a news item on our ZNBC TV on 24th May 2013 that Zambian Breweries has opened a new plant in Ndola which will be using cassava in addition to other commodities as a raw material in brewing of beer. This to me is a very huge milestone for the cassava growers in our country. We might not need to look to other countries for a market in cassava. This is happening in Mozambique by the same company as well as the newest state in Africa; South Sudan. I have doubts whether Mozambique produces more cassava than Zambia. The challenge now is with the ministry of agriculture and livestock, especially the departments of extension and marketing to organize the farmers into cooperatives so that they can meet the required volumes and quality demanded by the market.

Zambia Breweries being a serious profit making business venture would not want to buy the commodity the way our loss-making FRA does. They wouldn't want to be dealing with a 'cadre' of farmers. Cooperatives and SMEs should be encouraged to sign supply contracts with Zambian Breweries. As change agents, we should act proactively to knock at their door instead of waiting for them to come to us because they have so many alternative sources of raw material. They have taken a step to open the factory close to our high areas of cassava production to cut on transportation costs. This to me is a very big success story for the cassava industry, however, it will only be successful if farmers grasp the opportunity with both hands. It is a well done job to Zambian Breweries! It now remains with the farmers to do their part by meeting the required volumes and quality of the product. Cooperatives here would very important in making this market linkage a success. Allow me to knock at the doors of serious cooperatives in Zambia.

Cooperatives can be vehicles for economic development

This article was published in the Times of Zambia on 27th April 2013, and what I was trying to communicate to my readers is the poor work culture that has been developed in our cooperatives in this country. In other countries, such as some Scandinavian countries including the United States of America, cooperatives have been a very important 'vehicle' for economic development. Cooperatives in those parts of the world have ended up developing into well renowned banks, companies and factories just to name but a few. One great example is the Rabo bank of Netherlands.

May I express my amusement by the number of cooperatives found in our country! When I asked for a list of cooperatives in Eastern Province from some partner organization in some ministry, I was surprised to find that they are in thousands! These cooperatives are in various classes, being; enterprising, emergent, non-enterprising and dormant. The cooperative movement in our country has not been as successful as probably in the Scandinavian countries where we could have copied the idea from. In those countries, some cooperatives own some of the renowned banks that are even operating in Zambia and other developed countries in the world. The cooperative movement has not been successful in Zambia due to many factors and standing out as the major contributing factor is political interference. Not long ago, many Community Development members were being asked to form groups and cooperatives if they are to access developmental funds from the government. I remember the community department in the Ministry of Community Development spearheading the formation of these groups, and one past minister was distributing cheques just before the tripartite elections in 2011. This unfortunately, has been the trend for a long time, as long as politicians discovered that it was a great tool that they had to use to enhance their selfish ambitions of getting or staying in power perpetually. Very rarely is this money accounted for.

However, I boldly should state here that if we continue with this attitude or culture, our cooperatives will remain as they are today. The idea of forming cooperatives is to spearhead development to a group of people or households with a common goal. I should be quick to state that there are however, some cooperatives that are operating as they are supposed to be – going concerns. I know of one or two cooperatives that are operating in Southern province. These cooperatives are in the dairy industry. What has outstandingly struck me is that these cooperatives have managed to even employ a qualified veterinary doctor to offer extension services to its members. The cooperative buys dairy products from its members and sells the milk to Parmalaat and the other milk is transformed into various dairy products such as cheese and yoghurt. This is a very good example of how cooperatives should operate. Each member of the cooperative that supply milk makes not less than K5, 000 and the highest sale is around K24, 000 per month. If we had even twenty percent of all cooperatives operating like the cooperative in question, we could have reduced the unemployment burden to a single digit by now and our per capita income would be better than that of Botswana or South Africa.

Unfortunately, the bulk of our cooperatives operate like 'babies' meaning they are formed for the sake of accessing fertilizers and seed from the government. Immediately they get the inputs, these cooperatives stop to meet and close their shops until the next farming season. It is amazing that the bulk of them were formed by politicians who wanted to use them to remain in power in perpetuity. As a poor country like Zambia, we can't afford to have a cooperative that operates on socialist ideologies. I am alive to the fact that we are a Christian nation and that we should be helping the underprivileged, however, if cooperatives are the vehicle that we shall use to create jobs and help develop this country, they should be allowed to operate as socio-capitalists. Cooperatives should acknowledge that operating viably would enable them to make business profits. As a starting point, the Ministry of Agriculture and Livestock should deregister all those cooperatives that exist to just access fertilizers. I know the Minister of Agriculture and Livestock will be very unpopular but it will just be for a year and the benefits will be long lasting.

Let each cooperative have viable strategic plans of how they will operate and bring in resources. If the Mazabuka cooperative has done it, what can prevent the cooperative in Ndola to be successful? What twists my mind is that the same members of the cooperatives keep on getting the inputs and very few of them have ever built a meaningful storage facility to keep their produce. Year in year out, these members get fertilizer and sell it to commercial farmers, and some of the fertilizers even finds its way into neighboring countries. Secondly, responsible authorities should seriously think of reviving the Cooperative Bank. The bank will be important to give loans to cooperatives and once this is done, even banks that are risk averse to lending to smallholder farmers will join in because their clientele will dwindle if they don't do so. We need to make the cooperatives to operate as going concerns and not bottomless pits that are only revived in the planting season.

'Zoning' the country

In 2015 we were at the cross roads again in as far as agriculture is concerned in Zambia. We had a drought again; we have never had a big drought like that since the 1992 one when we all tasted the yellow maize which was famously called the Scott mealie-meal imported from America. It is inevitable that we were going to import some maize that year to supplement whatever we had produced. However, farmers rose to the occasion and produced more than we could consume and sold some to our neighboring countries such as Zimbabwe, Malawi, DR Congo and many others.

The traditional maize producing areas in this country are Eastern, Southern and Central provinces. We have some districts like Mpongwe on the Copperbelt, Isoka and Mbala in the Northern provinces and Kawambwa in Luapula Province that have become part of the maize producing districts off the maize belt. Nonetheless, the traditional maize producing zones had in the last three seasons become drier and drier and yet, our agricultural policies have continued to support and promote these areas as maize belts. We had a drought that year because of the effects of the El Nino. We saw the delayed onset of the rainy season, with rains starting in the second half of January in Lusaka for instance. However, areas in the north had reasonably favorable rains since the onset of the rainy season.

With these changing weather patterns, we need to strictly relook at the type of agriculture that we can promote in each of these areas. For instance, we know that crops like sunflower, beans and to some extent sorghum do not need as much water as maize. In most cases during normal rainy season, sunflower and beans are normally planted in January. In areas like Musungu of Kawambwa and some parts of Luwingu, they grow beans twice in one rainy season. We need to deliberately start promoting the growing of sorghum, cassava, sunflower, beans and rearing of cattle more in Southern, Western, certain parts of Central and southern parts of Eastern provinces more than maize. It will not help us in any way by providing maize seed, herbicides and fertilizers more to these areas and yet the crop does not grow to maturity because of inadequate moisture/water. If the Copperbelt, Northern, Muchinga, Luapula and Northwestern can produce as much maize, this can be redistributed to deficit areas that might not produce adequate maize. These are times when we can use more of the push than the pull strategies. However, what this will entail is that we need to ensure perfect market conditions for the crops that will be pushed on to these areas otherwise, people from those areas might end up starving to death if they produce sunflower for instance, and yet there is no good market for that crop.

Most of us think that it is difficult to make people cultivate crops that they are not used to. Indeed, many have been forced to think that maize is the only food crop because of the support and publicity it has continued to receive from governments in Africa. In Zambia, not long ago, our staple crops used to be sorghum, millet and cassava. In areas like Luapula for instance, cassava has continued to be the staple crop and these areas have never had hunger for a long time as compared to other predominantly maize producing areas. As a matter of fact, it has never occurred in many years that the government distributing food in Luapula because of inadequate production of maize. It is areas like Southern, Eastern and Central provinces that have continued to be victims of food shortages because they have been 'brain' washed to believe that maize is the only food. Areas like Western province have continued to be perennial recipients of food from government because of the floods for those people have continued to live on the banks of the Zambezi river; this has not been the case with people from other districts such as Kaoma that have adopted cassava as their staple crop. As we think of implementing our seventh national development and national agriculture investment plans, we should strictly rethink our policies and allow the technocrats to be drivers of such plans. The country needs to be zoned into areas of comparative advantages for each commodity for targeted efficient production.

Corruption, a thorn in the agriculture sector

Corruption is a vice that is found everywhere; USA, Europe, Australia, Asia and Africa. Many people think corruption is only an African problem but that is not true. However, it seems to be more pronounced in Africa because of the lack of political will to fight it by many governments. In other countries like America and Britain, immediately a public officer is suspected of being involved in corruption of any kind, that officer will step down to allow for the law to takes its cause because if they persist to hold on to their offices, they will be suspended. In our little country of Zambia, corruption started a long time ago but it became more pronounced after 1991 because it was strongly embraced by the leaders then. Very few Zambians that claim to be very rich got their riches through genuine channels; a lot of them that have mansions, companies and bath in unlimited riches had links with some controlling officers in government or people with vantage positions of authority in public offices.

Well, as the scope of the book is agriculture and because I am a son of poor peasant farmers; I want to highlight what corruption does or has done in regards to the cost of production and how it inversely affects job creation and ultimately economic development. The case study of course is Zambia. Transparent International defines corruption as the misuse of entrusted power for private gain. It is the dishonest or fraudulent conduct by those in power, typically involving bribery. A lot of people think corruption only takes place in public offices, no! Even in private sector, this ugly animal has thrived. Take for instance; a person that owns an input supply company and the government floats a tender to supply a particular product. The input supplier happens to know the key person who can influence the results for offering of that tender and he goes in the 'night' and negotiates with that particular person that if he influences the results, he will have a cut in the spoils. That is a pure case of corruption and that is happening in this country and the region. It is no longer a secret that there could have been corruption in the supply of oil tenders as revealed by the various commissions of inquiries instituted by the president in 2011. The effect that corruption has on agriculture is that it makes the cost of inputs higher. If corruption is involved in the supply of oil products, it increases the cost of producing a kilo of maize and that cost is transferred to the consumer along the value chain. What it means is that if one wanted to grow two hectares of maize, that person will be forced to reduce it to one because he can no longer afford the costs involved. This means that the number of people that would have been employed had two hectares been planted would reduce. This will also have a bearing on the efficiency of producing that kilo of maize. If that person could have produced it at K0.50 without corruption being involved, he would ultimately produce it at K0.80 because of the hidden costs due to the vice for instance, and this will go on up to the point the maize meal is consumed by the household down the value chain.

It is prudent that this topic is included in this book because the country has been subjected to a lot of strange decisions such as when the minister of agriculture (which mostly supports maize) was announcing the FISP tender results. It was indeed surprising to learn that through the revision of the tendering procedures in 2011, billions of kwacha was saved. The question many were asking then was how could the fertilizer be cheaper that year than the previous one when the price of oil which is a major component in transportation of the fertilizer had been raising? People were asking whether we had suddenly become so efficient in our operations that overnight we were able to cut costs? Many still felt that the tendering for FISP should be subjected to further scrutiny because there were still too many grey areas. The country has continued to suffer because jobs cannot be created as would have been the case due to the high cost of production and many that cannot cope are dying of hunger in the country because they can't produce.

It is worth mentioning here that the late president His Excellency Michael Sata need to be saluted for allowing the Anti-Corruption Commission (ACC) to investigate his ministers in regards to the allegations by some quarters of society that they were involved or influenced the results for the supply of oil and some works at ZESCO[14] then. The people that had been accused were very close colleagues to the president and it was really a big test to the genuineness of the fight against corruption. Many presidents that served after Kaunda had opportunities though to take a leaf from the good policies that the first president had put in place where he did not allow his ministers to be involved in 'public business'. He asked those that thought was good at doing business to leave public offices and venture into full time businesses.

Two months before publishing this article in the newspaper, I had received a call from one of the poultry farmer who wanted to know why the cost of animal feed was too expensive in Zambia although we are self-sufficiency in maize and soybean bean production which are the major raw materials. The answer to this question is partly answered in this article; there are a lot of unnecessary costs that are associated with the cost of production in this country and this was complimented by the revision of the minimum wage[15] at the time. China has managed to develop at the rate at which it has because the cost of production in that country has been low for a very long time. People will appreciate my point that even if China is a major producer of groundnuts and other agricultural commodities for instance, it imports

---

14 The cases have since been dropped although public outcry has been that there was a strong influence from the political masters

15 In 2011, the government revised the minimum wage from K250 to K500 per month

a lot of such commodities because they believe in value addition. They prefer exporting finished goods as compared to commodities because this is the only way it will give them more value in addition to creating more local jobs for the people.

USA, which is one of the most developed nations in the world, has been surpassed in terms of economic development due to the low cost of doing business in China. Zambia also need to check all the grey areas in terms of agricultural production to revive and stimulate agricultural production and encourage value addition. The country doesn't need rocket science to conclude that there could have been underhand methods in the supply of inputs for FISP for a very long time which needs to be cleaned up because they are making production inefficient in the sector. We know that corruption kills and it should be stopped for all especially in agriculture sector. Once corruption is minimized in agriculture, the country will notice significant changes in efficiencies and enhanced agribusiness ethics.

Follow ethics in crop production

The 2014/15 farming season was a very critical one as the country was hit by adverse effects of El Nino weather patterns. From the way the rainy season started in 2014, it was eminent that indeed some areas of the country were going to be worst hit while other areas were to receive too much rain which led to floods. Either of the two scenarios is detriment to crop production and what that means is that most crop commodities were to be in short supply. In mind are vegetables as these needs regular watering for them to grow well. This weather phenomenon led to a situation where the country had inadequate tomato supply and price of the commodity at Soweto market K300 ($50) a box (20kg). Farmers that had a crop that was about to be ready but has heard that the good price of the commodity and had just sprayed some pesticides on their crop whose residual period could be seven days for instance, had to be tempted to harvest the crop and ripen them artificially so that they can make money in the case by selling them.

Some will argue that there is nothing wrong the farmer had done as they merely wanted to take advantage of the high prices to make money as farming is a business. Such a farmer in my opinion is as bad as a thief or a murderer. Pesticides have a residual period in which one must follow before harvesting. Certain export markets such as Tesco in the UK which have strict minimal amounts of residual pesticides that can be found in food before they can be accepted would ban such supplier from being its vendor. For instance, certain pesticides should not exceed any quantity as low as 0.01%, and anything falling above this will be rejected and suppliers risk losing supply contracts. The reason why they have such restrictions is that certain pesticides don't easily get metabolized or breakdown in our systems and what that means is that with time, they accumulate in the body. They will not kill someone just there and then but a considerable accumulation over a period of time if not checked may lead to health problems and complications. No one will want his or her best customer to have healthy problems or complications. What this means is that one is just killing his or her market. Business that is focused on making money today without thinking of how it will be sustained in a long time is as good as dead.

It is against this background that farmers are implore to follow strictly what is written on the label of various pesticides that they use in the various production of crops. Sometime back, a good colleague of mine in the village lost two family members because it was suspected that they had eaten vegetables that were sprayed with a pesticide and did not follow the preharvest interval as indicated on the label. Initially the whole family started to vomit just after the meal and unfortunately, two of the youngest family members succumbed to death. Some scholars defined ethics (business) as a form of applied ethics or professional ethics that examines ethical principles and moral or ethical problems that arise in a business environment. It applies to all aspects of business conduct and is relevant to the conduct of individuals and entire organizations. Therefore, as farmers if we are not following the crop production ethics then we are not following business ethics because farming is a big business. Quoting Sallie Krawcheck (Global Leadership Summit), she noted that 'if it comes down to your ethics versus a job, choose ethics because you can always find a job.' I agree with her because it takes so many years and time to gain integrity. Integrity is something that cannot be traded off with prospect of making money just because there is a short supply of that commodity. Zig Ziglar noted that the most important persuasion tool you have in your entire arsenal is integrity. This is because people will normally doubt what you say but they will definitely believe what you do. Mind you consumers are indeed statistics while customers are people – they are watching what you are doing. The absolute fundamental aim in marketing is to sell or make money out of satisfying customers' needs, this is because if you get into a profession because you think you can make a lot of money, you can never become successful as you might be preoccupied with making money and not satisfying the customers' needs. Business is all about satisfying customers need sustainably. I found the behavior of the farmer who harvested vegetables that had still some active pesticides in it criminal. Of late, some multinational companies have done away with very good products because of the same problem of residues – that is acting responsibly. I know of a product we used for controlling termites called Chlordane. This product is banned the same way they have done with DDT. Some organophosphates have also been discontinued in the EU and America because of the detriment effect on the environment including people.

Therefore, as we get to rainy season every year in this country, we normally experience shortages of tomato, onions and cabbages. As business people that want to sustain the market so that we can supply even so many years, we need to observe production ethics in agriculture. Pesticides are quite safe but if the users are not following instructions, they can be harmful to us and the environment. As users they need to sure that they read the label three times before application and always indicate in their spray book the dates they sprayed their pesticides, that is, good record keeping is always important in business. Crop production ethics should be the norm in farmers' businesses. This is because 'if you can't describe what you are doing as a process, you don't know what you are doing'- W. Edwards Deming. Business ethics are always important and need to be followed to the latter even if we are competing.

The need for competition

It is a public secret that Zambia enjoys an agribusiness comparative advantage over its neighbors in the region. The country is blessed with a favorable climate, good environment and relative potential 'available' market for most of the agro-products. This is a country that has a youthful population with about sixty percent of the population under forty years. Certain economies in the developed world are now grappling with an aging population (the baby boomers). However, having potential is nothing if we can't transform it into the competitive advantage to attain the competitive edge. We should hasten to mention that we have done relatively well as a nation in the last decade or so though it is not good enough for the potential we have. The areas that the country is lacking so much are technology, energy and infrastructure including marketing infrastructure. Most of our agriculture is practiced in the rural areas that lack the basic infrastructure like roads, communication (internet, phone, railway, and air), manufacturing and others. We have scored quite well in the last twenty years to try and open up the rural areas. To everyone's satisfaction, this has of late (2013) seen an influx of investments in the agriculture and trade sectors. We are now talking of having about twelve or so fertilizer companies, eight seed companies, over twenty agrochemical companies with so many traders, over ten processors, so many commodity traders and several transporters. This is very good for our agribusiness industry. I think in the region; we could be second or third to South Africa in terms of development of our agribusiness sector. However, we all know that forty percent of the water found in SADC countries are in Zambia. The country has four very big lakes, five large rivers, so many lagoons, swamps and streams. With all this potential, it is sad to learn that only one percent of irrigable land is utilized. The country only boast of Nakambala as the largest flood irrigation with Mkushi farm block being first in overhead irrigation in the country. This is a huge opportunity needs to be explored if the country is to attain real competitive edge in crop production. There is no reason why we should only become so busy with farming in October and become docile certain months of the year; we are losing money.

Though we have encouraged good investment in input supply, the country hasn't done well in secondary and tertiary industries. It's good to learn that we may have three more companies in the fertilizer industry, two agrochemical and seed companies before the end of the year. I am saddened to see Zambian agribusiness companies fold up because they fail to compete on price and quality from products coming from South Africa. We have not perfected our agriculture well because of some factors mentioned in this article. I remember in one of my articles this year, I wrote that the first time I saw very good looking potatoes was at Johannesburg fresh market in 2007. However, Buyabamba and its out-growers are able to match that quality because they have access to good inputs such as seed, chemicals and many others. What I don't know is the average yields attained. However, we can't see investment in secondary industries such as those cutting the potatoes into chips ready to be fried. We still see some fast food companies importing chips from South Africa. Many of us have been compelling the government to ban the importation of chips from South Africa. The only alternative we have not provided is where the fast food companies are going to be buying already processed chips ready to be fried. The only way is to avail finance so that these companies can set up such upstream industries in that value chain.

When Hon. Sampa was the deputy minister in commerce, we were told that one firm in Ndola was setting up such a factory but it has since gone quiet. We definitely need competition in agribusiness industry if this sector is to tick. Therefore, let our trade advisors at most of the embassies and commissions worldwide sell the 'potential' in the agribusiness sector. We as Zambians should also jump on this trek so that we invest in agriculture especially after the energy sector is sorted out. We can invest in cloth making factories unlike exporting lint to China. May I please ask the MD for Dangote cement that the next time he is in Zambia; I will be pleased to meet him; he is one of the few practical Africans around. I know he has set up a tomato processing factory in Abuja and he is negotiating to set up a fertilizer manufacturing plant in Nigeria. I like his practical approach to business; no wonder he is the richest black African. Well done Dangote, you can still do more in Zambia though there is more you can improve on. Healthy competition is indeed very important in any sector to encourage investments. Of late we have seen great investments in construction and agricultural sectors. Zambia is the next investment frontier especially if the country can effectively fight to minimize corruption and continue to improve on infrastructure status. However, the country has over a million vehicles on its roads and all of them are imported as finished products. There is need to establish or re-establish in the case of FIAT motors the assembly plant. This will make the final products affordable. This is not only true for the motor vehicle industries, but all other industries as well. We have seen it with the cement sub sector and the companies are all operating as going concerns. This country is a hub for investments.

Zambia as the regional hub for agro-investments

Not very long time ago, many people used to wonder as to why most of the multinational companies involved or supporting the agricultural sector had been shunning investing in Zambia. Many of the multinationals either had offices in South Africa or Kenya and just used to sell their products and services in Zambia through third party companies and agents. They only established franchises with some Zambian or regional companies that had representation in the country. The disadvantage of getting products and services through franchises or third parties was that it made the technologies expensive thereby not getting the full benefit of such technologies. Just taking you a few years back in the 80s when we had FIAT motors with the assembly plant in Livingstone, most of the Zambians were affording to buying new brand vehicles from the showrooms. When such investments closed up operations in Zambia, we saw the inflow of second hand vehicles mostly from Japan which are imported through various neighboring countries' ports. The disadvantage of such procurement processes is that people buy goods or products that they have not physically inspected and many of them have ended up losing their money. The cars that are bought through such transactions may be faulty in certain instances.

In the agriculture input supply companies, Zambia had presence of companies such as DuPont having operations in this country before the MMD government come into power and thereafter, they relocated after the change of government. However, some of them relocated because of Kaunda's drive for nationalization. Amongst the first companies to come back after the reintroduction of multi-party democracy in the country was SeedCo. Later on, Pannar came on the scene and this created real competition as we had three seed companies unlike before when we only had ZamSeed. For instance, maize production was synonymous with seed varieties from ZamSeed such as the MM604 and other MM varieties. Later on, one of the renowned seed breeders in Zambia who once worked for ZamSeed established a seed company called Maize Research Institute (MRI) with some great seed varieties. This drive later, later compelled companies such as DuPont to come back as Pioneer and now we have Monsanto who are amongst the giant seed companies in the world. The local companies such as Kamano have also been resilient on the seed market, which have gone into production especially with their open pollinated varieties. We had EPFC in Chipata who were also adding value to groundnut seed multiplication in the country but has since closed up. This is a great opportunity for investment as groundnut industry is slowly becoming a giant commodity again. Later, we have seen the coming on the seed market of many other seed companies both in vegetable and field crops; such as Syngenta, Klein Karoo and many others. This should not scare any other seed companies that are thinking in investing in Zambia because this can be used as production country to supply other regions in Africa that do not have the climate like Zambia. Great strides also are being done in the region such as the Funwe seed company, Demeter and other companies in Malawi. This is the trend even in other African countries such as Zimbabwe and Mozambique though the level that Zambia has achieved is more advanced than these countries.

In the crop protection segment of the industry, the story is not different. Zambian companies used to trade in generic products from China and India. Though in the early 80s, we had Shell which transformed into Cyanamid and later Crop Serve which was bought off by United Phosphorous Limited (UPL). Crop Serve was at some point was the largest crop protection solution provider and was distributing even for multinational chemical companies such as Bayer, BASF Dow and Makhteshin Agan. Amongst the Zambian companies was FarmChem which earlier was distributing for all other multinationals (BASF, Bayer, and Syngenta as well) before CropServe took over the franchise or they shared some product lines. The other Zambian companies on the scene then were CropChem, Plant AgriChem, Prime Cropcare and some other small companies operating as sole traders. However, Amiran set up base offering not only the crop protection products but seedlings, communication as well as irrigation equipment. Later, BASF established a representative office and later became a legal entity in 2015. In November 2013, Syngenta bought off MRI and started distribution it products through MRI as a subsidiary company while Bayer have continued distributing through Precision Farming though they have their own marketing staff on the ground. In the fall of 2016, Bayer announced the acquisition of Monsanto.

The benefit of having these multinational companies establishing their offices in Zambia is that besides creating job opportunities, we have seen some products becoming slightly affordable due to competition. This is good for the farmers because their production costs will ultimately start to go down and their commodities can be competitive. There are more companies that are eyeing Zambia as the best investment destination in the region. Not long ago, we saw Yara Fertilizers getting a stake in GreenBelt; the implication is the immerse competition to other fertilizer companies that have been operating in the country before. The investment has not just been in the input companies but we have also seen an influx of commodity traders such as Cargill, NWK, CHC, AFGRI, LDC and many others. This is very good indeed for our agriculture industry, however, the government through its parastatal institutions responsible for policy formulation and regulating healthy competition need to be alert as we are likely to see unhealthy competition as they jostle for business opportunities in a small market. We are likely to see some collusion or 'dog-eat-dog' competition especially before the cake can be grown bigger. This is a reality; some of players that have been long in the industry will agree that there are some companies promoting seed technology to the farmers than what we have seen in a few decades ago. Some seed companies never used to promote their technologies through establishments of demo plots but yet they used to win larger fractions of tender through the government's small scale subsidy (FSIP). How was demand created to the farmer? Where the farmer only dreaming of their presence or some push strategies were employed by some officers in some office somewhere? For the farmer, this investment by stakeholders is more than welcome and farmers have never been as excited as they seem today because of these massive investment inflows in the sector. All the government can hope is to see these giant companies invest into growing the cake bigger by investing in opening up of new production opportunities through financing some operations unlike fighting for the ten thousand hectares than the government is subsidizing the farmers. Zambia indeed is an investment hub and those companies that are sitting on the perimeter fence still thinking whether to go and invest in Zambia, need to make the move now! This all turn around of events has come about because of the consistence of the the country's policies on agriculture, the rapid development of infrastructure to support such businesses as well as the stable and predictable political environment that exits in the country.

The effective management of stress in crop production

In 2014 main production season, it was mentioned that the region was going to receive less rains than normal. Meteorologists (weather technologists or specialist) predicted that the country would have a weather phenomenon called the El Nino in Southern and parts of East Africa. What this meant was that the region was to have droughts or reduced rainy activities in the region. El Nino is a weather pattern that is disastrous to countries like Zambia which practice rain fed agriculture. Normally, our rainy season starts in November to April but today. However, during that year by month end of December, some farmers had not yet planted because they were still waiting for the rains. The unfortunate thing is that even commercial farmers in this country depends on rains to fill up their dams or replenish the ground water for those using boreholes for irrigation. During the 2013/14 and 2014/15 seasons, Zambia had been receiving reduced activity of rains especially with regions I and II. As a matter of fact, farming blocks like Mkushi had been planting about fifty to sixty percent of their potential irrigable land for winter crops.

Nonetheless, farmers can't fold their arms just because they are not receiving adequate rains to water their planted crops; they need to look at the best practices that they can follow to salvage whatever they can get from what is planted. Agriculture is their business. Firstly, this has to start from the time they plan what to plant. Generally, in times like we had, it is always good to diversify what farmers plant or enterprises involved in to spread the risk. For instance, they need to plant drought resistant crops such as sorghum and for those that go for maize, they have to go for early maturing varieties and those varieties that have in the past exhibited tolerance to droughty conditions. When cultivating in times of droughty conditions, it is always good to follow conservation agriculture from the onset. For instance, at the time of planting do not disturb all the soil except just where the seed is to be placed. This can be done by drilling lines where they are going to place the seed as opposed to plowing the whole field. In so doing, soil water is conserved as the moist soil underneath is not exposed to direct heat from the sun. Secondly, they need to leave mulch on the soil surface to act as a 'guard' against sunlight. Prior to planting, there is need to ensure that the soils are loose enough for health development of roots. Remember that roots are the foundation of any healthy crop. There are also some products on the market which stimulates formation of healthy roots.

The next fact that need to be observed in such instances is the need to be timely in application of fertilizer. In the case of basal fertilizer application, this should be done at planting if farmers can but in instances where that is not possible, it needs to be applied immediately the shoots are out. Additionally, weeds will be competing for the same inadequate moisture in the soil. Therefore, there is need to timely weed-off the weeds in our fields. As most agronomists will always emphasize, the best time to control the weeds is before they germinate (pre-emergence of weeds). Again, farmers may realize that their crop could be the only 'greenly' staff around, hence, the leaf eating pests will always feed on them. Therefore, it is prudent that the control of such pests is done immediately they are noticed in the field or better farmers could spray preventatively especially if they are using systemic products. Luckily enough there are some good systemic pesticides that can remain active in the crop for some time. The next factor to observe is the control of fungal infestation which leads to fungal diseases such as rust, powdery mildew and others. However, this might not be a very big problem during very dry conditions but farmers need to know that regardless of the conditions, they may have fungal diseases of some kind especially if crop rotation is not effectively practicing in their cropping pattern. Luckily enough, there are some very good fungicides currently on the market in Zambia that contain Strobys and Triazoles. The beauty of most of the Strobys if rightly applied, have properties which can help manage to some extent prolonged drier conditions than crops that have not been applied with such products. They also help the crop to efficiently utilize nitrogen making the crop remain green for a long time. Companies like BASF have trademarked this property on their Opera and Abacus (fungicide) as AgCelence. The Triazole part of the fungicide is the one that does the curative on the disease. For professional advice always consult specialists in agriculture as this is the best way to learn. Nonetheless, there are some very good technologies now that can enhance the farmers' yields.

Impact of technology adoption on productivity

In 2013/14 growing season Zambia cultivated about a million hectares of land under maize from about 1.3million small scale farmers.

From that production, the country only managed to harvest about 2.5million tons of maize, giving an averaging yield of about 1.9mt/ha. That was not a good yield in a country with a perfect climate and relatively fertile soils with abundance water. The question that seeks answers is why?

Zambia is an underdeveloped country. It is not that the country has no answers for the low productivity; if anyone from the agricultural sector is asked including the cleaner for the permanent secretary's office, they all have the answers by their fingertips. They will bring them out like bombshells one after the other if asked to. It is something that sounds funny but very unfortunate in that the country has missed opportunities to rake-in millions of dollars because its neighbors like Zimbabwe, Malawi and the traditional brothers from Congo DR had been having insufficient maize stocks since 2012. Zambia has very hard working farmers but yet they are not been improving their productivity to take advantage of the existing market opportunities that presents themselves in the region. Productivity has remained stagnant at 1.5 to 2.3mt/ha for the smallholder farmers who are in the majority for over thirty years since 1988. Some vague definition of insane suggests that it is the doing of the same thing and expecting different results. Some of the biggest challenges that smallholder farmers face in Zambia is lack of 'full' adoption of useful productive technologies. The country needs a complete shift in its peoples' mindset; for instance, it is not possible to effectively cultivate more than ten hectares with a using a hoe even if preparation of land is started in May because agriculture is not all about maize production. Smallholder farmers in this country seem to have a weird belief that herbicides destroy the soil. This is something that many technocrats have failed to understand even after demonstrating to farmers what technologies can do to productivity. In 2012 production season, I deliberately demonstrated some technologies by cultivated three hectares of a field for my old lady within two days; the first day was used to plant the seed and the following day we went in to spray with Lumax an herbicide used in maize combined with Gramoxone for controlling weeds that had already germinated. In that particular field, there were no significant weeds as most of them germinated when the crop was already grown. A lot of people in the village used to go and watch how the maize was doing although even though the crop was planted late on the 24th December because of the late onset of rains. The yields were relatively good in that we got an average yield of 4.85mt/ha while a lot of the fields around that farmstead were harvesting less than a ton per hectare.

The following season, the same field was cultivated and planting was done on the 28th December. A neighbor that admired the happenings in the field the previous season asked for the herbicide and gave him a litre of Stellar star to be sprayed in his field. It was so annoyed to learn that he was discouraged by his colleagues that the crop will not germinate if an herbicide is used gain; so he did not apply the herbicide. In spite of using minimum tillage by just drilling lines and planted the seed, the crop was amazing and that person ended up blaming his mentors who advised him not to use the herbicide. What was most annoying being that the product had even expired the time he wanted to use it; meaning the K400 that was used to purchase the product was just wasted. This scenario is not unique to that particular village or farmer only but it is the trend countrywide. As change agents, there is enormous work ahead of us to educate the farmers on the safety of such technologies. Farmers need to understand and appreciate the different types of agrochemicals and what they do. It is though important to maintain a health soil micro fauna but we should know which class of chemicals can have an impact on the population of the soil microorganisms.

Secondly, it's quite odd that someone who has planted 20kg of seed maize which is equivalent to a hectare would think of applying four bags of fertilizer and expect to get over 6mt/ha. In Zambia, farmers have been under utilizing fertilizer in the quest to economize without supplementing with organic fertilizers for instance. Additionally, farmers have unfortunately taken to believe that to successfully grow a crop, they only need seed and fertilizers as inputs; They need to understand the basics of farming from soil pH, nutrient status of the soil which is a function of the cation exchange capacity, water requirements and disease and pest control. Farmers also need to start appreciating terms such as soil compaction, soil humus which is a function of organic matter and water retention capacity. It is does not mean to say farmers need to be enrolled in an agriculture college but they need to learn to consult the technocrats just the same way people do when the courts find them wanting. Farming is not for amateurs. In productivity improvement, technology is just but one factor; access to finance is another big problem in sub Saharan Africa. Agriculture will be incomplete if technocrats and practitioners don't talk about financing opportunities.

Financing agriculture for productivity

In Africa especially sub Saharan Africa, productivity in agriculture is very low. In most of the value chains, average productivity is lower than a quarter of the potential. The bulk of the producers who contributes over eighty percent of the food basket consumed in the region are smallholder farmers. In Zambia we have over two million farmers that annually cultivates over one million three hundred thousand hectares of maize annually. This is not different to Malawi, Zimbabwe and other countries with an exception of South Africa that seem to have a high concentration of commercial farmers as opposed to smallholder farmers. As we have already discussed in this book, smallholder farmers are faced with an array of challenges which has led them to be so inefficient in producing. One of the contributing factors to low productivity is lack of access to finance for the to acquire modern technologies. This does not mean we do not have financial service providers in the region; to the contrary, there are so many banks in the country both local and international.

So why are the banks not coming on board in as far as providing financing to the agricultural sector? Smallholder farming is a big market opportunity for any bank. Indeed, banks there to make money but the money has to be made sustainably. In Africa and Zambia in particular, banks have concentrated on servicing only clients in formal employment and commercial farmers at the expense of the many smallholder farmers. Sometimes one may doubt whether they have done their homework in knowing the demographics of their potential clients in the farming sector. The bank that will develop products that are tailored towards offering solutions to the smallholder farmers sustainably will make money. As banks have concentrated on servicing those in the formal sector, some civil servants have been loaded with loans such that they are almost suffocating. Indeed, banks are into business to make money but business without spending is not business at all. Most of the banks in Zambia are behaving like 'shylocks' or are shylocks of some kind.

One is compelled to think like this because the smallholder agricultural sector is the single largest employer in this country; literally all the people in the rural areas are farmers of some kind but then they do not receive the service which is supposed to complete the jig-saw puzzle. This will lead to the fact that the agriculture sector in this country will not develop if farmers do not have access to finances. We all know that business is all about spending money to make money. Look at what has happened in 2015 in the tobacco segment in Zambia; the largest pre-financiers in that value chain did not participate in their core business and this led to a reduction in area cultivated of about sixty per cent. This is a very big loss in revenue because the industry is a source of forex as well as contributing to employment opportunities. In countries like Malawi tobacco is the single most earner of forex and it contributes to over seventy percent of the agriculture GDP. We know that towns like Kasungu were developed mostly from resources realized from the tobacco sub sector. This is not different from Zimbabwe as well. Therefore, no matter how much advocacy is done to promote the use of certified seed and other inputs, as long as the smallholder farmers don't have access to affordable financing they will not have the money to buy fertilizer, seed, chemicals and other inputs and implements needed to improve their productivity. The implications are that limited jobs will be created and their contribution to GDP will stagnate. This will also have great negative ramifications to the development of the manufacturing sector.

In 2015 every farmer in Zambia so excited when the government finally gave a go ahead to ZAMACE to implement the warehouse receipt system (WRS). This is one of the key risk mitigation factors for the banks to come on board because the output from the farmers can be deposited with merchants through ZAMACE as the regulator. That piece of legislature should have been accented to a long time ago but they say its better late than never. However, the country still has a big challenge with storage space but this is an opportunity which the 'Donald Tramps' of Zambia should jump on. If productivity can increase by doubling, Zambia can be harvesting in excess of 9.6million tons of corn alone. However, our storage capacity including those run by the private sector is less than four million tons and this is an opportunity which we all need jump on. With the implementation of the WRS in this country, construction of storage facilities is a business that we should all venture into. Banks need to come on board as they are a second partner in development after the government. We know that nothing can be done if money is not made available.

With such piece of registration now being an Act, the sector is assured of establishments of out-grower schemes and forward contracts with farmers and this will make commercialization easy as smallholder farmers will be involved into commercial markets of promising or lucrative value chains. The sector anticipates to see births of out-grower schemes in maize, soybeans, sunflower, vegetables, fruits, small livestock even fish. For a long time, everyone has reiterated that with the favorable climate, fertile soils and sheer number of farmers involved in agriculture, our crop productivity should double which ultimately will lead to achieving the first ever true bumper harvests in various value chains. As the government is revising the national agriculture investments program (NAIP), they need to set productivity targets that will make farmers sleep in 'gumboots' and make them happy at the end of the day. In the past we have seen targets that are only on paper and never implemented at all. Productivity improvement is attainable with the right strategies and support from our colleagues in the private sector such as the banks, commodity merchants, multinational input suppliers, transporters and the farmers. This should include the research, schools as well as training institutions such as universities and colleges in the faculty of agribusiness.

In a nutshell, the humble appeal to the banks is that they need to invest into smallholder agriculture because that is where the country has large numbers of participants and the major supplier to the national food basket. Though the government is trying its best with the FISP, they can also do much better by implementing certain good policies which are only on paper. Governments worldwide generally are very bad at doing business and the private sector should be the main drivers of productivity, however, they should create a real enabling environment for everyone to thrive.

Zambians will remember that since 2002 when the late president Mwanawasa introduced production subsidies in agriculture, the country has been a net producer of maize which is the staple crop. Since then, the economy had been doing quite well posting positive growth until later in 2015 when the prices of copper plummeted on the world market. It is evident that these subsidies will not continue in perpetuity due to limited resources. The reality is that the cost of inputs will be dear to most of the farmers and if not handled well, might start having challenges to even produce enough to feed ourselves. This will be very unfortunate because the country might just see some of the investments put in various value chains such as processing in soybean close shop because they will not have adequate raw material for processing. This calls for well thought out policies from the government through those tasked with the responsibilities such as the minister of finance and his team including those at commerce. However, with the filling in of the gap that will be left by government by the private sector such as the banks, we are all confident that we might just see better results than we have achieved as a country.

In this country, we have companies like NWK, Cargill, Zambia Breweries, COMACO, Illovo, Buyabamba and others that have been running various out grower schemes in various commodities such as cotton, maize, groundnuts, soybeans, sugarcane, barley and potatoes. These companies should be encouraged in that they are bridging a very important factor of production – finance, which under normal circumstances is supposed to be provided by financial institutions at affordable interest rates. In Zambia, securing financing from financial institutions is very expensive as interest rates are very high, ranging from thirty (30) percent to as high as hundred (100) percent in certain instances. It is no longer a secret that our economy which depends mostly on copper has of late (2014) not been doing well. Though late, this is the right time to look to agriculture as one of the main sectors that can turn around our misfortunes. The government needs to encourage these companies that are making great investments in agriculture at high risk by putting up policies which will make them invest even more in agribusiness. We all now know that a Zambian farmer when supported will produce and we can only create wealthy if we venture more into production than just being mere traders through retailing.

The plea is that other companies, especially those in value addition should emulate what their colleagues are doing if they are to remain relevant in business. There is normally a tendency of processors not investing in production to target farmers that might have been pre-financed by other companies by offering a few cents above what the pre-financier had offered at the time of contracting. This leads to side-selling and what the farmers forget if involved in such underhand method is that they are killing themselves. Though we know that outgrowing of commodities is a problem worldwide but where it has been managed and handled well with good supporting policies, it has worked well and removes the burden on government. With this in mind, I hope the finance minister and that at commerce can come up with deliberate policies to encourage out grower schemes. For instance, they could reduce company tax for any company which is pre-financing say a minimum of hundred thousand farmers in any value chain. If that is done; many companies will start pre-financing farmers with various inputs so that the farmers can produce. This will not only encourage the traditional farmers to produce but newly agricultural graduates that might not find formal employment as well as some week end farmers like many of us. Agri-financing is one area we have fared so badly in Zambia and once we get a saviour in this segment, all the idle land may become productive overnight and the country will realise the dream of being the grain basket of the region and Africa.

Lastly, farmers should change their mind set; the earlier they realise that they are no longer in the humanism environment the better. When farmers have the opportunity to be provided with inputs through pre-financing, they need to honour their contracts by selling back to people that helped them with the inputs even if there could be someone dangling some juicy carrots. They should remember that by pre-financiers investing in farmers without collateral, they exhibited confidence and farmers need to pay back to harness such relationships. The only way farmers can realize good return on investment is by putting to good use the inputs they have been provided with. In the past we have seen some farmers get eight bags of fertilizer and sell half of that to get money for '*kanchina*' (illicit beer) and expect to get two hundred bags in a hectare. With the right support and hard work, we can revamp agriculture and taking it to heights never seen before in this country.

Revamping the agribusiness sector

Agriculture is one of the oldest sectors in Zambia and world over. When God created the world; the first two sectors God concentrated on where Agriculture and Engineering. It is worth to note that agriculture must have been brought to light before man. Man, found the trees, fruits and animals already created. However, many of us have a question as to how has this sector stagnated for a long time when it is supposed to be the lifeblood for Zambia's economy? New sectors like technology are doing better than agriculture. Many of you who happened to be there before the third republic would agree with me that the Kaunda started well and agriculture was doing relatively fine. He had a vision in that he established secondary industries to tap into the primary products of agriculture in all the provinces; talk of Mulungushi to buy lint cotton through LintCo; Mukuba, Livingstone and Kafue Textiles to buy secondary processed lint; the City Clothing factories to be buying the finished products from lint; the pineapple processing plant in Mwinilunga to mop up all the pineapples produced by the people of that region; they were so many other industries like the Munushi Banana Scheme in Mwense; Kateshi Coffee of Kasama; Mpongwe Wheat Scheme in Ndola; National Milling and many others too numerous to mention.

Besides the good things that the third republic brought, courtesy of the MMD government; it killed all the above mentioned industries, save for a few that have been revived from the 'intensive care unit' like National Milling. The biggest mistake that government did was to undo everything associated with Super Ken as Dr. Kaunda was famously known. This was a grave mistake and Zambians to an extent are to blame including this writer because they allowed it. Just imagine if the country had decided to build on what the old man had started, we could have been a giant economy in the region and the continent as a whole. Well, as if that was not enough, the government then went on to 'defile' the agriculture sector by killing marketing and production units of the sector. Zambia was reduced to a market instead of a net producer in spite of all the potential that it commands in the region. A lot of terminologies were coined to hoodwink Zambians and farmers in particular about potential in agriculture.

The agriculture sector only started showing signs of life when the late president Mwanawasa took up the reigns as the republican president. It is gratifying to note that in the last four seasons (2004 – 2008), Zambia has been sufficient in maize production. However, I am one of the few that refuse to subscribe call it agriculture because agriculture is much broader than growing maize. It involves the integrated production of crops; field crops, horticultural, floricultural, cash crops, plantation crops and rearing of animals, birds and fish. It involves infrastructure development that supports production of all such commodities as well as good transport infrastructure including feeder roads, railway, processing industries that serves as markets for the primary products, storage infrastructure such as the silos; affordable and accessible financial services; the good agricultural education delivery systems such as the colleges, universities and delivery of effective extension systems; good policies that are not skewed to one commodity only but the whole industry. If we are to microscopically investigate some of these points highlighted, we will be shocked to find that only one area was favorably covered – maize production and marketing. By the way, having great respect for politicians I was taken aback with the finance minister's debate in parliament in 2012 when he said that cotton is not exported in its raw form. I know there are light moments in parliament but members should learn to be serious when they are debating issues that affect livelihoods and life; cotton is just cleaned and separated from seed before being exported. The three merchant companies in the country export lint in its raw form. No meaningful 'value' is added apart from just separating the seed as earlier stated which is later sold to farmers for planting in the subsequent season.

Nonetheless, I was delighted by the policy pronouncements and budget allocations done by his government in that budget year. If the 8,000km of road network that PF's government had earmarked to construct is something they will accomplish; the establishment of increased energy generation and distribution and support to establishment of cottage and processing industries as preached by Hon. Bob Sichinga is something to go by, then the country is definitely in the right direction to improve agriculture. It is not a secret that even if we produce a lot of agricultural commodities; these are not competitive in the region. The biggest costs come from production and transportation. If the country is to increase the power generation and distribution, this will spur a lot of proliferation of industries including those that are involved in manufacturing of agricultural inputs such as fertilizers and chemicals. When the road infrastructure is improved and new road networks opened up like in Lukulu's rice producing area; our rice would price competitively as to that from Thailand, for instance. It is true that where roads go, investments normally follow. Every meaning Zambian has no doubt that the steps taken by the government to develop infrastructure if fully implemented, will favour the development of this favourite sector called agriculture, and when the agriculture sector flourishes, there will be no need to preach about job and wealth creation because these will automatically come. Folks, just like the late president Sata, I feel ashamed to be eating tinned beans from South Africa when forty percent of water in the SADC region is found in Zambia. What is wrong with the country producing and tinning its own beans? There is no pride in being a begger, and Zambia begs for things that the country can produce easily.

As a country we should be ashamed to be importing energy from South Africa; or worse off learn that Botswana is self-sufficient in beef when quarter of that country is dry (desert). It is disheartening to further learn that our only hope of exporting tinned beans and tomato paste to America has been shuttered by the receivership of the only industry we had. I had been praying day-in and day-out for the leadership that they continue with what they have started. The country can see some dim light in agriculture after a very long time. The only hope and talk on everyone's lips is that the private sector such as banks, merchants, traders, input suppliers and many others can come on board to support wholeheartedly these strides to revamp agriculture because business development belongs to the private sector and not the government.

The private sector as the driver

In October 2012[16], USAID in collaboration with its partners held a conference to discuss some of the interventions needed to develop and grow the agricultural industry especially in third world countries like Zambia. The conference was quite educative and brought out so many pertinent issues in the industry. This is a conference that drew participants from world over. During lunch of the first day of that conference, I sat with a colleague from Cargill and we brainstormed some of the issues that really need to be done in Zambia to forge ahead with agricultural productivity and production.

However, what had impressed me that particular week was the good news that Cargill had invested over US$2million in setting up a maize milling plant in Chipata. When I first heard the news, my heart jumped with excitement such that had my mouth been open, it would have landed in my hands. The reason is that the Zambian government made some positive pronouncements some two years earlier that it was going to invest in establishing milling plants in each and every district across the country. That pronouncement was made by Bob Sichinga in the famous industrial cluster sermon in parliament when

---

16 October 2012

he was the commerce minister. There was nothing wrong with that statement but when one critically analyses the implications, that was taking us back to the times of INDECO. As much as the government meant well for the farmers in the country, it was taking two steps forward and three backwards. The reasons being that the government is mostly, not efficient in running such profit based businesses. What would have happened or what is mostly likely to happen is that they will employ people unnecessarily and whatever will be realized will go into overhead costs and other non-productive activities. Secondly, just think of it having over seventy-three milling plants including one established at Bwalya Mponda at the heart of Lake Bangweulu where the main commodity is fish and a bit of cassava production. The idea was excellent but the way they want to execute it is not sustainable in modern societies. Putting myself in their shoes, I could have opted to do it differently.

All that the government needed to do was to engage private sectors like Cargill and entice them to establish plants such as what they have done in Chipata. In my opinion we don't need more than two to three states-of- the art milling plants in Eastern province; such investments should grow with productivity. The other negative connotation government investment into maize value addition has is that they will be defeating the purpose of crop diversification. Just imagine the value of investment which will go in establishing over that number of milling plants regardless of the time frame. This will affect other value chains like soya, cashew nuts, groundnuts, fish, livestock and many others. What the government needed to be do was just to 'point' their fingers that let there be a milling plant in Chipata, and the partners run to establish it like Cargill had done. However, that normally comes with good predictable policies with investments into public infrastructure like roads, water transport, communication and energy in place. In no time, the country could have seen all the heavy-duty vehicles causing congestion on Cairo road move to Mtenguleni, Chadiza, Lundazi to mop up the grain and lint in that part of the country. What Cargill has done is commendable and they should receive 'kudos' from the government and every well-meaning Zambian.

The country need to see more of such investments being established in other rural parts of the country like Musungu in Luapula province, Mbanga in Lukulu and many other areas that are struggling in the area of commerce. Though Cargill has a very strong foot print in Eastern province, this is a sign indeed that they are positioned themselves to stay in that part of the country and sooner than later, they will attain competitive edge over their competitors. We only hope that other partners' players in the private sector emulate what Cargill has done. If we have much of such investments in the rural areas, the country can be assured that productivity will improve because the plant will demand for more raw materials/commodities. I know other partners like BASF, MRI Syngenta, Greenbelt and others that offer extension service besides agricultural inputs have positioned themselves by already establishing depots in that province. That investment will make them work hard to increase production through improving productivity by offering better production technologies to the farmers. What a good innovation and heed to government plea by Cargill! The onus is now on farmers to produce more to feed into that investment so that it is not a white elephant. Indeed, productivity improvement must be tackled with the private sector as the key partner and driver of growth!

Covering up for the lost opportunity

One day in the middle of the rainy season in 2014 as I was taking a tour of selected farms in Southern province, it was evident that some fields had received less than five good rainy days. In many fields, it hadn't rained for a long time and the crop was wilting. Factually, during that season, some parts of the country had effectively received only two months of rainy days and many of people thought the rains were going to prolong a bit. Many farmers were still planting as late as mid-January. They were planting medium to early maturing varieties for maize and soybeans with a view to harvest something. That year, most of the seed companies had ran out of early maturing varieties for maize (the 400 series) hence, farmers were forced to plant either the 600 or the 500 series out of desperation.

That particular season I planted my small field as late as 26th December with MRI 595 and under normal circumstances at this time of the year, this should have been planted with MRI 455 on such a date but Iike everyone else, I was desperate as there was none on the market. The gamble was that if that crop could receive at least two to three rainy days in March, it would be enough to see it to maturity. However, knowing that some weather forecast in the region had predicted chances of having El Nino and this weather phenomenon comes with droughty conditions while the opposite is true. Nonetheless, I was comforted by our local weather pundits who predicted normal to above normal rainy season. It was never a normal rainy season though. Many crops looked as though they had a perfect grain filling but the kernels were very light. The rains went at a critical time when the crops needed it more and to make matters worse, it even become too hot such that the little moisture that the soils could harvest in the voids evaporated. Farmers lost an opportunity to make money that particular season. So, what should they have done?

In the earlier chapters in this book, we discussed strategies of applying conservation techniques in agricultural production and some of the benefits that accrues for adopting such strategies. Those that did follow smart agriculture during that particular season lost some yields but not as bad as those that stuck to the conventional agriculture – they are wept! What then do farmers need to do in these uncertain times of good weather patterns? Farmers and extension service providers need to look at what other opportunities are there in times of bad weather so that they can still make money in our businesses and advise the farmers accordingly. Suppose people tried horticulture; that is the growing of vegetables. To venture into this segment of agriculture, farmers need to prepare well in time. If they wait until August or October in the case of Zambia and other southern African countries, they should not be surprised if they fail even find enough water to raise a nursery. The early bird catches the worm or Bemba's would say, *'akabangile kanwine ayalengama'* (meaning the same thing). Farmers need to start their enterprises very early and seek professional advice from crop specialists to ensure that they utilize effectively every drop that comes their way. They need to aim to ensure that every twenty litres of water utilized should give them not less than 10 kg worth of produce. When it is so dry crops like maize tend to have certain diseases. For instance, in that particular season it was observed that there was an outbreak of diseases on the summer crop such as *Diplodia* on certain seed varieties. This disease last affected maize some good eight years earlier. There was also an outbreak of many pests such bollworms on green maize. As expected Diamond back moth on cabbage was a menace worth mentioning, red spider mites in potatoes and tomatoes, as well as unprecedented outbreak of aphids including the yellow ones in sugarcane were very common. This was because there weren't a lot of green material for the pests to feed on and the crop was the only source food available. However, most of these are pests that can easily be controlled if farmers bother to use some of the very good products currently on the market in Zambia. Though, many farmers cried of certain herbicides that failed to control their weeds in their fields that season. There are so many factors that we could attribute that to.

Lastly, as a Nation there is need to endeavor to invest in equipment that can provide the farmers with more precise weather forecast. I do not get much value in some weather forecast provided by our experts; it does not make sense to me. Forecast statements such as *'there will be cloud cover in Luapula and thunderstorms in places.* The question I always ask is which places are those? Can't the weather man be a bit more precise with the areas like our colleagues on CNN, BBC even South Africa does? To be frank, many farmers would rather they don't broadcast such forecast. In this era of farming, farmers need to cultivate with resilience if they are to remain in business. It is also become evident that some of the pesticides that used to effectively control certain pests are being phased out because of their negative impact on the environment. Chemical companies that have remained in conventional research and development of pesticides that have a negative bearing on the environment will soon be out of business; people are now conscious and looking for biological products for instance.

Cultivating for 'resilience'

For too long now we have seen farmers complain that they had a crop failure because they did not receive enough rains. If they are asked how much they could have received in that particular season, some would say 500mm and above. This is more than enough to see a crop of maize to maturity. Many might argue that what matter is how the distribution was – indeed, distribution is important because it is possible that this could all have been received in a month or two and we know that a maize crop with the current varieties on the market will need not less than 90 to 150 days depending to mature. However, the problem of distribution has not started in the last decade or last year year. This has been a problem for a very long time but farmers seem not to cope or learn lessons for this unfortunate weather pattern. Someone once told me that insanity is doing the same thing and expecting different results. Why have we been doing the same thing and expecting different answers?

In modern agriculture, every drop of the rain or irrigation matters. Don't allow any runoff if you are practicing irrigation because not only are you losing water but soil and nutrients too. We need to be reducing ourselves to small living microorganisms that are in the soil in order to understand better what is happening there. Farmers should learn to think and plan resilience in agriculture if they are to ever get on top of low productivity which has been with them for a very long time. There is no way a farmer in Mpongwe can be producing an average of eight tons per hectare while those in Masansa or other areas are getting a partly 1.8Mt. There is no way a farmer in USA should be getting 15Mt when he uses the same seed, fertilizer and implements. Only if we have such a mindset will we ever win this battle. Mind you it's only less than fifteen years away before the planet will have an extra one billion mouths to feed from the same area of land. If farmers are struggling to feed us now, how will they manage to feed an extra one billion people?

The answers lie with all of us as actors in the agricultural value chain. There are only a few basics that farmers need to appreciate if their productivity has to improve; these are plant material, water and soil water holding capacity, disease and pests, and soil pH. Conservation Farming Unit (CFU) has been training farmers about conservation or smart agriculture; they have been telling farmers that it is possible to get the best yields if there is less disturbance to the soils. What they have been indirectly telling us is that we minimize the disturbance to the soil; soil organic matter will build up in our soil. This organic matter (OM) is important for water and nutrient retention in the soil. The habit of turning up the soil during plowing has a negative impact on the buildup of organic matter. Additionally, plowing to the same depth year in year out tend to allow for the development of a condition known as a hard pan in agriculture, especially if farmers are not occasionally ripping their fields. Once farmers have sorted out the land issue, the next thing to take care of is the soil pH. Different crops have different pH ranges from which they perform better; forcing a crop to grow in soils with pH less or more than its optimal will NEVER give them the yield they desire. The majority of the crops though will perform better in pH range 5.5 to 6.8.

Farmers also need to ensure that they have the right planting material; for instance, it will be folly of them to plant 700 series if the onset of the rains have delayed or if they are found in Southern province and yet they do not have irrigation to supplement their production. Farmers need to know the material's potential and characteristics, and then manage the weeds. Sometimes, seeing some fields planted near the main road, one wonders whether some farmers are propagating weeds. Weeds are thieves that should not be tolerated and farmers need to be advised about the best time to control weeds. It's like a child, the best time to mold him/her into a good citizen is when they are still young, don't wait until they have tried some awful habits.

We all know that crops just like animals and human beings are living things. They do suffer from diseases and are attacked by pests. I remember when growing up in the villages, we used to have diseases and sometimes even lice if we don't maintain our hair properly. Crops do suffer from 'malaria' and they can sometimes have 'lice'. The malaria in crops like maize for instance, are fungal diseases such as grey leaf spot and 'lice' would be things like aphids and the famous army worms.

Luckily for farmers of today, they have very good products on the market which do not only combat these diseases but they prevent them and in addition, these products have a combination of more than one active ingredient. In most cases one the active ingredient of the two present in these fungicides and of most importance are *strobys*. These help the plant to use nutrients efficiently as well as protecting them from diseases. Farming today is not just about producing a crop but how cleverly and efficient farmers produce them. Resilience in production is the in thing which separates agriprenuers from crop producers.

Is agribusiness on the right track?

Just before the closing of the marketing season in 2014, I received a call from a concerned citizen who wanted to know whether agricultural development in Zambia was on the right track. Wanting to know other people's views, I posed the same question at him. He gave several answers, some of which I never even thought of. I appreciated that readers' views, he was so knowledgeable and I wish he could share those with the policy makers that are driving agriculture in this country. I had my own views which could be right and or wrong; it depends on which context one is viewing them from. Let's discuss some of them in this book although many of them could have been shared already in the previous chapters.

There are always two sides of the same story but let's start from the positive view. For those that have closely followed agriculture from the first republic, they will agree with me that our agriculture has taken a bias towards maize production. In most instances in Zambia, when we talk of agriculture we are most likely talking about maize production save for our friends from Southern province that integrate with animal rearing. In one of the Sunday interviews on ZNBC TV when a former minister for Southern province was featured, he boasted that at some point in his agriculture endeavors he used to produce more maize than a combined production from Luapula and Northwestern provinces. Indeed, some farmers have mastered so well the art of growing maize such that in a difficult year like 2014 was, Zambia was the only country in the region with a maize surplus. The country had recorded maize production on the increase from several hundred thousand tones some few years back, to a couple of million tones. In the same vein, the country had also relatively recorded remarkable improvements and increased production in other relatively new crops such as soybean. About five years earlier, soybean was only grown by commercial farmers but now it is a crop grown by many farmers. Some people used to refuse to eat the crop and claimed that its products smells but today; the country is getting most its cooking oil from soybeans. Additionally, other food items such as milk, the famous *nyamasoya* (soya chunks) and many other products are being made from soybeans too. The country is seeing areas that never used to rear cattle valuing the importance of cattle; such provinces like Luapula and Northern provinces who when they see such animals, the first thing that rung in their heads was meat.

Between 1992 to about 1999, our productivity for maize had slumped to a ton per hectare but currently, it is slowly rising to between two and two point five for the small-scale. Additionally, the country has recorded remarkable investments from large multinational input suppliers such as Syngenta, BASF, Bayer, Amiran, NWK and commodity traders too. Five years earlier, it was unheard of that companies would provide leverage funding to small scale farmers; they always wanted collateral first. Today, as long as a farmer has a field and growing either maize, soyabeans or any other crop, he can walk into an agro-lending institution such as AFGRI, NWK and get a loan for a tractor as long as he was a member of a cooperative though the products are limited. We have seen some input suppliers providing short term credit facilities to small-scale farmers which was only privilege of commercial farmers. This is a right step in the right direction and we can only hope that the farmers will not disappoint as was the case with Lima Bank some years back.

However, on the other hand, one still feels that the productivity level of less than three tons for maize is not good enough as well as less than a ton for both soybeans and groundnuts. Additionally, the financial institutions need to come on board and provide financing opportunities products tailored to this sector in the small-scale market segment. They have concentrated in providing loans at expensive rates to the people in the formal sector of which most of them ends up buying assets such as cars that are not as productive. Banks are killing each other to fight for clients especially those that work in government and quasi-government institutions, unfortunately some are using even underhand methods. Bad or unethical marketing strategies such as 'dog-eat-dog' competition is being used. Nonetheless, despite these great achievements in agriculture it beats me to see a very weak manufacturing sector in the country which is not supporting the production sector. The manufacturing sector is supposed to be absorbing all the raw materials churned out of agricultural production. Only a few companies are seen to be involved in processing or value addition such as ZamBeef that have taken a complete value chain approach which is very good for Zambia's agriculture sector and job creation strides. Actually, in the agribusiness sector, it is only Zambeef that has really taken the challenge seriously, though we are slowly seeing companies such as Cargill following suit. If you go to Zimbabwe, you don't find some of these chain stores that are here in Zambia and making profits on our heads at the expense of the locals. The Zimbabweans owns their economy. Its everyone's wish to see a very strong agribusiness manufacturing base in Zambia. In fact, the country started quite well under the leadership of Dr Kaunda when we had companies that were producing products such as Tarino, Kafue Chitenge, Mwinilunga pineapple cannery and many others. The country is slowly waking up when it comes to power generation; there is no way industries can invest in manufacturing when they have enough power to run small industries. ZESCO has been insinuating that it would be too to invest in solar power. The company needs to rework its cost benefit analysis and just. The country, I believe is on the right track but what is worrying most Zambians is the pulse at which we are moving. Everyone knows that agriculture is a business and it is a profitable venture indeed. We can share and prove the profitability of agriculture.

Is agriculture profitable in Zambia?

A business is any activity, enterprise, firm or company in which goods and services are exchanged for one another or for money. Important to note is that in any business there is a form of investment and customers to whom its goods and or services can be sold to in order to make a profit. With this definition, agriculture qualifies to be a business in Zambia because people make investments in form of inputs and there are people and firms that buy the outputs. Emphasize should be made that it is a business elsewhere too! If the environment is anything to go by, agriculture is supposed to be the biggest industry in Zambia. This is because we literally have everything that is needed for effective agricultural investment; in business terms, the country enjoys the comparative advantages than any other country in the region. The industry is supposed to be contributing more to the gross domestic product (GDP) than even mining.

There has been debate as to whether agriculture is profitable in this country. Straight to the point, agriculture in Zambia is a profitable business depending on how it is run. Just like any other business; its profitability depends with the level of business planning, investments and execution. In this country, there have been some farmers that started from a scratch and have made a fortune. On the other hand, there have been others that have failed tremendously. Normally this is how any business is. The biggest challenge with Zambia's agriculture is that the players sometimes choose the wrong enterprises to venture into. They rarely conduct our market research so thorough. Take for instance; some farmers have normally ventured into certain enterprises after hearing how successful other people have been in some enterprises. They have normally done this without daring to understand and study the challenges such people could have faced. By the time, they tried such ventures, there could have been so many people growing that crop, hence the low prices or the ropes of success could have shifted a bit and such have ended up being 'bruised' in business sense.

However, one of the biggest problems that has been noted in agriculture which farmers normally have not done well after the market research is the actual production itself. Many times farmers could have rightly understood the market but rarely planned of how their production will be done for their produce to be competitive. Farmers have always wanted to produce the crops with the least inputs and expect to get the best yields they have ever of.

In business they say 'garbage in, garbage out'. You cannot expect to get the best of yield from soybean seed for instance if you used recycled seed. Unlike when you use fresh seed bought from a recognized distributor or the producers of the seed, there will be likelihood of achieving eighty percent of such yields. Secondly, some farmers have bought certified seed, fertilizers however, they have waited until they see the weeds coming out before they start to weed. We have always emphasized in this book and on several interactive platforms that the best management of weed can be attained when it is done before the weeds germinate. Mind you, the greatest loss that farmers get in their production comes from poor weed management. Weeds are thieves or 'plunderers' that are not supposed to be entertained in farmers' fields at all costs. If they can manage their production and have reasonable yields, such as at least having to achieve over seventy percent (70%) of the potential yields, our commodities will be competitive; they will attain a competitive edge serve for other factors like infrastructure. Have you ever dared to wonder why soybeans from Brazil are cheaper than our own soybeans grown at Chipata when sold at Kasumbalesa or millers in Lusaka? Or why the maize from USA is cheaper than the maize from Mkushi or Choma? Many of us will think it is because production is highly subsidized in those countries. That reasoning is partly right, but the biggest reason is that the farmers from those countries have perfected their production techniques by using the right inputs at the right time, and the support systems such as government policies favors them. Suffice to say agriculture is a profitable business if we apply ourselves properly as business people. Nonetheless, there a lot of constraints as compared to colleagues in the developed countries but those challenges should be opportunities that farmers need to harness and understand well. Fellow farmers, agriculture is the only business which is renewable. Mining is an industry that is exhausted; you can't go back after years to mine the same minerals in the same environment. Some of you that have lived in Kabwe will agree with me that Kabwe used to be an economically vibrant town before 1993 and all was because of a well-developed mining industry. If one dares to go to Kabwe today, you will be very brave if you will not shed a few tears – it is a sorry site especially the famous mining compounds. Agriculture therefore, is a profitable business if we dare do our homework. There are so many tools that farmers can use to aid them make decisions. Remember in the early chapters of this book, some tools such as enterprise budgeting was explained; that is one tool farmers can use. Due to inefficiencies in certain value chains in Zambia, the government implemented a ban importation of cooking oil. Was it a good decision to make?

The ban on cooking oil importation

On Wednesday 25th March 2015, I was listening in to a program on one of the local radio stations which quoted a former permanent secretary for Luapula that he blaming the government for having banned the importation of cooking oil. His argument was that Zambia does not only have enough crushing or processing capacity to satisfy the market but he further went on to lament that the country did not have enough feedstock of sunflower and soybeans to supply the industry. It was rather an amazing if not perplexing statement coming from the former senior civil servant. To be honest with him, I chose to disagree with those un researched 'facts'.

Since 2010, Zambia has been a net exporter of soybeans. Before then, the country used to import soybeans and most of the oil products from soybeans and sunflower. In 2009 or earlier the government banned the exportation of raw soybean and the idea was to allow the extraction of cooking oil for the local consumption through various value adding activities. That decision worked to the farmers' advantage in that it offered a ready market for the soybean. This policy led to the establishment of added crushing capacity by companies like Zamanita and the country saw the birth of Mt Meru and Emman on the Copperbelt. In 2011/12, while working on a Bill and Melinda Gates Foundation funded project promoting the development of the soy value chain, many farmers were recruited to venture into soybean production. A number of small scale farmers responded to the call to diversify their production because a market was readily available. By then, the importation of oil from palm oil which was coming via COMESA countries from Malaysia through east Africa was banned. The reason they banned the importation was that it was cheaper than our locally produced oil. My supporting this decision should not be misconstrued to be promoting inefficiency here but this was still a young industry and it needed support for it to mature.

The former PS might argue that by promoting restriction of oil importation, the country was promoting laziness; not at all. These oils are coming from economic giants that have perfected the competitive advantage over years; the farmers definitely can't compete unless the country is ready to sacrifice this industry and forget about job creation. These decisions are not synonymous with Africa but even well developed economies in Europe do sometimes venture into such activities. Producers should not be in a hurry to get a good life when they haven't worked for it yet. The country would rather we buy our expensive cooking oil now and develop our own industry so that with time, we will improve the efficiency and compete with any similar product on this planet. Actually, at some point even the export of raw soybean was effected. Now, if the argument from the former permanent secretary hat the country did not have adequate stock of soya and sunflower, why in the first place did the country allow export of the 300,000 MT of soybeans that same year? That decision by the minister was supported by many Zambians because the country had adequate crushing capacity and raw material for cooking oil production. There were only a few individuals that could have been benefiting from that trade. As a matter of fact, the country's crushing capacity siting with Mt Meru, Emman, National Milling, Tiger Feeds, Zamanita, Gourouck, even SMEs such as Mucha Enterprises, the Kalumwange Women Group, Musungu Enterprises in Kawambwa, Katete and Chipata Women Development Associations should be more than 500,000tons.

If feed stock is the problem, our farmers had just started to harvest their soybeans (April 2015), sunflower and groundnuts. So there was no need for the former PS to worry at all. What he could have waited for was the determination whether the country did not have adequate raw materials, then that was when government should have allowed the importation of sunflower and soybeans. This country need jobs and those jobs can only be created by such interventions as promoting value addition; the treasury need taxes and such can only be collected if the tax collection is widening through promotion of value addition capacity in SMEs. The ban on the importation of cooking oil was done rightly because the country today is self-sufficient in cooking oil production and it is a net exporter of soybean products to the region and other by- products such as soybean cake.

Nematodes; a menace to crops

Some five years ago one of my nephews suddenly started losing weight even though he was eating normally. The parents took him for various tests at clinics but he was not improving. However, after trying several medications, one of the doctors asked the parents when was the last time they had dewormed the child? Since he had just come from the village, they had no idea whether he was dewormed at all. However, after being given deworming medication, we were all amazed at the number of worms that came out of the boy's small tummy. Thereafter, the boy started gaining weight and reduced on his 'bad' eating habits.

This type of 'poverty' is very dangerous because it has the potential to go unnoticed. This scenario is similar to what happens in the soil. The soil besides nutrients, comprise of so many microorganisms; some of which are useful to plant health and growth while others are harmful. Amongst the useful micro living organisms are those that help to decompose the dead matter. However, there are some that are harmful; in which category are found the nematodes. Nematodes are worm like small living organisms that are found in the soil. The devastating root nematodes is scientifically called *Meloidogyne species*;

there are a hundred of such species. There are of two types; the parasitic and the free living. The harm done on crops is similar except to state that it is very difficult to infer if the most prevalent nematodes are free living – both compete for nutrients with our crops. The parasitic are easy to notice in that if one uproots a plant which is infested with parasitic nematodes, there will be nodule like malformation on the roots. If not controlled, they can lead to the wilting of the plant and premature death. Examples of such are normally found on plants like tomato, cucumber, tobacco, egg plants and other crops of this family.

On the other hand, the free-living nematodes will live near the surface of the plant roots; normally in the area called the root zone. These will not infect the plant but they will be competing for nutrients with the crops. The damage caused by such nematodes is not as severe as the parasitic but they can significantly reduce the yield and in severe cases cause crop failure in that the plant will not access adequate nutrients for effective growth depending on the population. Plants may exhibit signs of nitrogen deficiency as well as other nutrients even in instances where you applied adequate fertilizer or nutrients. These will normally affect even crops such as maize, soybeans and most vegetables. These are what are termed the 'silent burners' in agriculture because they need an experienced eye of an agronomist to deduce what they are. In developed countries, there are laboratories where you take soil samples to know the population and type of nematodes found in your field. We do not have such facilities in Zambia yet and this is an opportunity for investment. So, what can we do to protect our crops from those 'micro thieves' of nutrients?

Thanks to globalization and technology transfer, we have got some very good products on the market that can minimize our nutrient loss to nematodes. For the free living, we seem to have very few products that farmers can use especially for non-transplanted crops. However, treating your seed with systemic insecticides may help. However, for transplanted crops like vegetables (rape, onions, tomatoes, cabbage, and so on) including other non- vegetable crops that can be transplanted like tobacco and non-transplanted like potatoes, we have very good products on the market like the newly introduced Solvigo from Syngenta. I should salute the team of scientists that came up with this formulation of Solvigo; it is a very good and an effective nematicide that farmers have used on my vegetables with so much success and have appreciated it. The beauty about this product is that you only apply it once and will protect your crop from nematodes up to the point of harvesting. As if that is not enough, it also protects your crop from early sucking pests such as aphids, whiteflies, jassids, thrips and other sucking pests. This product does not percolate into the ground water and can be applied to all soil types. If you are growing tomato, eggplants, tobacco, vegetables, potatoes; this is a product not to miss. However, this should only be applied between three to five days after transplanting; do not apply when your crop is fruiting. However, there are other products like Oxamyl that you can apply days after transplanting but they are not as effective as Solvigo. Bayer also has a very good product for nematodes called Velum. It has gone like wild fire especially with the tobacco farmers in the region. If you checked your soil pH, applied the right amount of fertilizers and have been watering your vegetables adequately and have been applying fungicides but your plants are wilting or yellowing; think twice, you might have nematodes in your soil. Nematodes are like tapeworms in the tummy of a child; do not tolerate them! We can conquer nematodes the same way Shaka did with his cow horn formation.

Preserving 'money' in the harvest

Whenever it is the harvesting season in relation to the agriculture industry in Zambia, a lot of things comes to the minds of the farmers. Many things are on their minds but the two outstanding ones are; wasted 'bumper' harvests for many of their crops and mostly maize, and low commodity prices. For the period 2006 to 2016, farmers had been blessed with good harvests. However, roughly more than thirty percent of what they normally harvest goes to waste. So many factors may be attributed to this including inadequate storage infrastructure, sheer poor planning, lack of knowledge and carelessness in rare instances.

Nonetheless, in the last few seasons, farmers have indeed looked after their grains quite well though they still incur some post-harvest losses especially at household level. It is unfortunate to lose ones harvested crop after having worked so hard to produce it.

On the market, nowadays, there are so many good products that can be applied to one's grain to keep it free of weevils which are the commonest pests in stored grains. One effective product that farmers have been using for some time is Actellic Gold Dust; it is a very good and effective product for grain storage. All that one needs to do is to shell off the grain and sprinkle this powder onto the grains before thorough mixing them and putting the grains in the bag. If my memory serves me right, one needs only to apply an equivalent of matchbox full of this product to a 50kg bag of grain. The maize grain will remain intact without being attacked by weevils for a long time. However, they have a period in which they will need to come and reapply depending on how they have stored the bags. If one does not want to come and retreat the bags, they can be spraying the outside bags with another product called Actellic liquid every fortnight to prevent new infestation from attacking their treated grains in the bags. Once this is done, the farmer is assured of eating pure white nshima in December as was the case in May when the grains were just harvested because they would not be weeviled. This is also a hedge against losing monetary value of the grain if it has to be sold later in the year.

This was not the case way back in the village, during the time we were growing up as we could have pure white mealie-meal up until November and thereafter, almost all the maize grain would all be infested, and because they could not break each and every grain to remove the weevils that normally lodges itself inside the grain; they would mill the maize with the weevils. The mealie meal and nshima would look darker than normal and not as appetizing as the one before regardless of the relish on the table. Apart from probably benefiting indirectly from the protein of the weevils, the meals where never enjoyable. Some mother would be heard complaining that mealie-meal made from maize that was highly infested with weevils never used to last long. Therefore, there is no point in the twenty first century to subject out children to that torture.

Lastly, may I take this opportunity to alert certain farmers that have the habit of buying aluminum phosphide tablets that they should refrain from that habit is that is a very dangerous product. That is very dangerous which is supposed to be handled by only qualified and licensed fumigators. Aluminum phosphide should only be handled by trained people that have fumigation licenses. If the tablets are baited in the bags as some people wrongly do and the bags of maize are kept in the house where they are living or sleeping, they risk killing everyone that is sleeping under that roof because that product produces a gas called phosphine which is a very dangerous gas that kills anything living. I have seen rats, snakes and other living things including ants die when they have been trapped under those tarpaulins in which they are fumigating. If people do not die from the gas, they risk causing fire because when those tablets come into contact with moisture, they explode into a fire. A certain commercial farmer lost tons of his grains when he did not cover their grains well which came into contact with moisture during fumigation. For people that don't have a license to fumigate and have not been trained, aluminum phosphide tablets are no go area; such people should stick to products that they can easily use and are less harmful to themselves and the environment. With these information farmers should be equipped and they should not be expected to lose even a single grain of maize, rice, soybeans and many other grain crops due to weevils/*impese* infestation. Farmers should understand that grains are their goods; that is where they get their incomes. Therefore, they need to preserve their goods. Additionally, they feed the nation including presidents; and they should value their job which is the best in the world. The weevils can destroy our harvest just the same way the army worms can lead to loss of a crop.

Fall army worms in Zambia

"There is an outbreak of army worms"! "No, they are stalk borers, but they look like army worms!". These were some of the few statements making rounds with the Zambia's farming community in 2016/17 production season. The Fall Army worms (FAW) attack on last production season devastated crops and they significantly reduced yields for the farmers in many African countries including Zambia. This was the first time this country had an outbreak of the pest, FAW. So, what are FAW?

These are pests scientifically called *Spodoptera frugiperda*, of the Lepidoptera order. They are said to be native to the Americas. The first infestation in Africa was reported in Nigeria in January 2016. They later spread to Southern Africa exactly one year later in January 2017 including Zambia. These pests are known to thrive well in the tropical environment. In such conditions, they are so prolific and breed continuously with about six generations per year. The adult moth lay eggs on host plants in clusters of 100 to 350 eggs The eggs have a white, pinkish to light green color and spherical in shape and hatch in about three to five days. The moth lives as an adult for seven to fourteen days and the full life cycle is about thirty-six to sixty days. They are known to be nocturnal chewing pests. In Ghana, the outbreak led parliament to declare an 'agricultural state of emergency' in 2016. Unlike the normal army worms, we have been used to in this part of Africa, the FAW cause extensive damage to both the leaves and the stalk. The chewing of the leaves reduces the photosynthetic area and they may also cause structural damage due to feeding in the whorl and this can result in lodging of some crops such as maize due to cutting of the stems. In some instances, they may direct damage to the grains due to larvae feeding on the cobs.

As highlighted farmers had in this region of Africa had challenges because, firstly, they could not know whether the infestation was army worms or stalk borers. It had to take some time before the entomologists declared that the infestation was FAW. These pests can be known by a few distinct features that can help the farmer identify them. It has on the eighth abdominal segment some four dark spots at the rear. It also has a broad and pale color along the top body with dark stripping at the sides as opposed to the normal army worms. On the head, they have dark net-like pattern which is oriented upside-down with a white 'Y' marking.

The pest can attack maize, sorghum, rice, wheat, sugarcane, cotton, groundnuts, tobacco, soybeans though gramineous plants are the preferred ones. It is this robust nature that makes it survive from one season to the other. As if this is not enough, it likes feeding in the night and during the day it will normally burrow in the stalk and hide there from the heat. Once the larvae are in the whorl or are inside the cobs, control is very difficult. The pest builds resistance to insecticides very quickly and it is prudent that we rotate insecticides if we are to win the battle against this menace. It is not only important to rotate insecticides based on the active ingredients but on the mode of action. Now that we understand and know the feeding and life cycle of the pest, how can we control it? Last season, this pest really troubled the farming community and it is likely to be back again because it has overwintered and survived in winter crops that are grown in the country. The country fared so poorly during that season in controlling the pest because of not following the right advice from the technocrats; for instance, it was learnt that the pest hides in the whorl of maize during the day and most of the country's public extension staff were recommending pesticides such are cypermethrin, Lambda and others which are contact insecticides. Both these pesticides are pyrethroid with similar modes of action. The government procured these pesticides at huge costs and distributed to farmers but nothing significant happened in as far as control is concerned. Maybe had drenching of the solution in funnel been recommended, the results would have been different.

Well, army worms have always been there in agriculture. We hope these will not be given a name as African army worms. If we can all bother to remember, Zambia had an outbreak of army worms in 2011 season and mostly Eastern Province was the one badly affected. As earlier alluded in the text, these insects undergo four stages depending on where you want to start from. Firstly, they have butterflies or moths which lay eggs and the eggs hatch into caterpillars which eat the greens and once they mature, they bury themselves and become pupa. The pupa will remain dormant in the soil until conditions are favourable for them to become moths and lay eggs again. They can remain dormant in the soil for many years. This is also true for other worms such as the stalk borer. In 2017 season's outbreak, the country had both the fall army worms and the stalk borers. It is important to try and understand the science and feeding characteristics of these two types of pests if farmers are to effectively control them. Army worms eat the leaves of the crops while the stalk borers mostly feed on the contents of the maize stalk; that is, they bore through the maize stem.

The army worms are the easiest to control because they can be controlled by contact insecticides while the stalk borers can better be controlled mostly by systemic insecticides but if one choses to use the contact, they need to pour the solution in the funnel of the crop so that there is contact with the caterpillar inside the maize stalk. The other point we need to understand is that there could always be an outbreak of these worms every year but they do not reach the threshold; thereby, there is an increase of pupa in the soil every year and when the conditions are favourable for the pupa to turn into moths, then we have an outbreak. The best we need to be doing every season is to be spraying preventatively for them. This applies to all pests in other crops. The next calamity that we are likely to face with the smallholder farmers is the outbreak of fungal diseases in soybean such as rust and powdery mildew. The reason being that we have seen a good number of smallholder farmers adopting soybean production but none of them is using fungicides to control diseases. This is leading to a build-up of fungal spores in the soil and crop residue. I hope when we have an outbreak of such diseases soon, the Kachepas will not blame it on the breeders or the opposition as the case may be.

By the way army worms cannot survive in herbicides as some have been suggesting that these could have been spread by application of these products. Remember army worm eggs are living things. What we need to know is that each army worm moth is capable of laying between fifty to several hundreds of eggs which hatch within two to four days and they are quite versatile in that they continually feed and reach the mature stage in just a couple of days. Products like Fastac (alpha cypermethrin) will easily control them. Therefore, do not insinuate that there is anyone trying to sabotage the efforts of the farmers or the government; it is nature and the weather was just favourable for the outbreak.

The other anomaly observed was that the bulk of the pesticide that was procured in 2017 by the government for onward distribution to farmers were mostly generic. This is not to insinuate that generic products don't work but the source of most of these generics and the quality were questionable. In case of an outbreak of the same pests this season which is most likely the case, farmers should not dare use the same products because they may be helping the pest build resistance to this class of pesticides. There are other products that are effective though they may look slightly expensive at face value. There is a product called Hunter and that has worked well including the control of Tuta absoluta in tomato. This product was a savior to most farmers that used it. There was also another one on the market called Proclaim fit. These are R&D branded products. Also, it was noted that a combination of Emamectin Benzoate with Acetamiprid or imidacloprid saved a lot of farmers' crop. The other thing farmers need to practice is by early planting of their fields to escape the infestation and also by planting early and practicing crop rotation. FAW is a disaster and farmers need to use tested scientific methods of controlling the pest. The good news is that the outbreak is in most parts of Africa and multinational crop protection suppliers are up in arms to find the lasting solution for this menacing pest.

Should we invest in marketing or productivity improvement?

For a very long time now, Zambia seem not to have found the right formula of how to deal with agriculture especially at smallholder level. The country's agriculture of the first and second republics was highly subsidized by the government end-to-end. However, in the third republic as earlier indicated in this book, there was an economic drive to liberalization which led to major institutions that supported the agricultural industry privatized; such entities like Lima Bank which was important to providing agro-financing and NAMBoard that was providing inputs and the market for the farmers were all privatized and recklessly disbanded. This had its own pros and cons to the industry. The role to improving productivity and providing market opportunities for the farmer was left at the mercy of unstructured market forces. If asked to offer an opinion about what the right policy for agriculture should have been, a balance of the two would have been the most appropriate until such a time that the private sector was fully functional.

The agribusiness sector is still in the infancy in this country including certain countries in the region especially where productivity, diversification, marketing, value addition and agro-financing activities are concerned in the sector. These factors need to be balanced and should work to complement each other if the industry is to develop. Nonetheless, favoring policies that will put the private sector to be the key driver of the sector especially in the marketing activities is the only sustainable solution. It is a well-known fact worldwide that private sector are the key drivers of economic development. It's a fact too that governments subsidize the agricultural industry heavily including developed countries like the United States of America to create what is known as an enabling environment for the private sector to thrive. The biggest challenge that we have in Africa and Zambia in particular is that in spite of having the comparative advantage on the natural resources such as land, we have inadequate technologies and access to affordable finances to realize the potential we have. As a country we should not always be begging for money from developed countries; but technologies that will enable us to attain higher productivity to enable our goods compete for international markets.

If the government wants to subsidize agriculture, they should concentrate on intervening at production end of the equation by creating enabling environments. The household level productivity needs to be improved in all the value chains in agriculture. This does not mean the government should just sit back and watch the private sector 'rob' the farmers when it comes to marketing; but they need to play a regulatory role through established quasi government as well as private institutions. There should be enormous investment in improving production efficiencies and this can only be done if the private sector is encouraged to own the bigger part of inculcating technology in our farmers as well as being the drive in creating marketing opportunities through embracing value addition. This are some of the opportunities that have not been fully tapped in Africa. Nevertheless, there should be fostering of strong partnerships between the private and public institutions. The government should encourage the private sector to invest in productivity improvement while it concentrates to invest in social sectors such as education, improving the rural infrastructure such as roads, energy, dam construction and communication. By education, we don't mean just attaining secondary education but going a step further by identifying certain skills in people and harnessing them. The day the private sector will own the extension service delivery in this country; productivity will improve over night. Additionally, the heavy intervention of the government in the marketing activities really crowds out private participation. There should be a balance between the two.

The government's role should be that of creating an enabling environment and strengthening private sector organizations that are involved in regulating fair play in marketing such as the Zambia National Farmers Union, Grain Traders Associations and many such institutions. Developing policies that will favor the private sector participation in marketing and moving into rural areas such as investing in value addition should be their main role. The policies should be consistent to allow the private sector have confidence. This has been the biggest challenge for this Patriotic Front government though some positive interventions have been noted. At some point, many thought they were not ready to be in government and thought the still were recovering from the opposition hangover. Like many scholars have said at different fora, the day the private sector will own extension service delivery – productivity will improve and the marketing service provision – markets will no longer be an issue. As things stands now, farmers are always at the mercy of the government to rescue them especially those that have been stack with maize cultivation. This does not mean the government should fold its arms and watch akimbo, as private sector commits all sorts of unethical business activities. They should be stronger players and partner with the private sector but their roles should be to ensure that the inputs are landed at affordable prices at the door step of the farmer. Therefore, the question of whose role is it to improve productivity and provide market opportunities in agriculture may be answered by the business community. The authors view is that the government's role is to create an environment that will be easy for the players to conduct business sustainably though to some extent some processes should be owned government but the private sector must be the drivers.

The high prices of mealie-meal

In Zambia the government has been subsidizing both the production and marketing of corn. Since 2013 when the maize subsidy was removed by government the prices of mealie-meal have been on an upswing. This has raised a lot of concern with the population as maize is a staple food crop in Zambia. It is not still clear as how maize had overtaken crops such as cassava, millet and sorghum. As a young boy I remember vividly having sorghum as our staple crop and maize was only intercropped to be eaten as green and for incorporation in beer brewing activities. As late as 1986, sorghum was the main crop grown in many villages in Zambia. The good guess is that trouble started when maize received a lot of support in terms of research, marketing and input provision by the government. The country is now at the crossroads because the crop that has received so much support seems not to be favored by the changing weather patterns especially in the traditional growing belts. There has been prolonged dry spells during the growing periods of the crop and this had negatively affected yields. Entrepreneurs have not helped the farmer here because very little has been done to promote smallholder irrigation knowing that over eighty percent of the farmers that grow the crop are smallholders.

In Zambia and the region as a whole, the farmers are always complaining about the high costs of inputs and the low commodity prices. On the other hand, the consumer complains about the high cost of mealie-meal which is normally the end product in maize cultivation. This is normal in any business because naturally a business person is 'greedy' and wants to always get abnormal profits. In the case of maize, mealie-meal will never ever be cheap if farmers do not learn the ropes of improving their productivity. It is important to know that a products pricing starts at the time the farmer is planning to grow or produce it; this is the case with mealie-meal. Actors in this value chain needs to ask themselves such simple questions as; what is mealie-meal composed of or how is it made? The questions sound basic but they are very important in this value chain. We all know that mealie-meal is the final product that comes from tilling the land, planting the seed, weeding, irrigating, applying fertilizers, harvesting, transporting, milling and marketing the final output. If businessmen understand the cost implications at each of these levels of this value chain, then only can they effectively solve the paradox of high cost of mealie-meal in Zambia. It's worth mentioning here that this become a political commodity in most of the countries in sub Saharan Africa. Politicians in many countries have lost power whenever this commodity becomes so dear to the people. The simplistic and most basic way of solving the problem is to start with improving productivity. That is all that is needed! We have always emphasized that regardless of how good a parent is; he/she will only bring out a responsible child if discipline starts at conception. They cannot have a responsible child if they were cheating at the time of conception. In short, we need to work on reducing the costs of production and once this is done, then will we have moved a step in improving productivity. This is because many farmers are growing the crop, even any other crops without putting the right amounts of the required inputs. There is no way the price of mealie meal will go down if the farmer is yielding 1.8mt/ha when even if they have planted a hybrid seed if they apply less than eight bags of fertilizer. On average, the cost of fertilizer alone for maize in a hectare is around K2,200 using the 2014 prices. Without adding other production inputs, the rough estimate of pricing implies that a farmer should not sell a bag of maize below K62; any price below this will mean that the farmer is making a loss (this is before adding the cost of plowing, harvesting, packaging and others). However, the same cost at the average yield of 5mt/ha gives a minimum price of a bag of maize to around K22. Even if the farmer was to add thirty percent of other production costs, a price of K45 per bag will be profitable. It is only when the cost of maize; the raw input in mealie meal production is low, will the cost of mealie meal come down.

Therefore, the onus is on every stakeholder in agriculture to ensure that productivity in this country is improved while the costs of inputs are maintained or go down. This can be achieved for instance by improving distribution through development of infrastructure especially in the rural areas where production takes place. On the cost of inputs, it involves every person including the police officer manning the road block. For instance, if officer asks for a bribe to allow an over loaded vehicle to pass at the roadblock, then that cost will be recovered from the fertilizer the transporter is delivering. The whole back falls on the consumer because ultimately, it is the consumer that has to bear or pay for all the inefficiencies along the value chain. To start with, there is need to ensure the farmer is educated about good agronomic practices so that he does not 'propagate' weeds instead of growing maize. There are a lot of ways to control weeds; mechanical, chemical and many others. If farmers can have weed free fields which are well fertilized and without any infestation of pests, then they will be on the right track to achieving their desired yields of 6tons/ha and more. Lucky enough, we have the magic and 'medicine' for the farmer to have that desired yield in Zambia. Once the farmer does their basics well, there will be no need of the consumers to cry to government to introduce the consumption subsidies. People should know that subsidies that are introduced to lower the cost of consumption are a drain on national resources. They do not add value to the industry but makes it inefficient. The government need to be promoting subsidies that will have a multiplier effect on production. Therefore, if the status quo continues, the people will keep crying for the high cost of mealie meal and it must be made clear to them that mealie meal will never ever be affordable especially to the common man. The prices of mealie meal can go down, therefore, if farmers improve their efficiencies. This decision should never be left to politicians though they need to play their role of policy formulation which will create a conducive environment for prices to go down.

Late in 2015, the government procured solar powered grinding mills were installed in rural areas and operated by Zambia Cooperative Federation (ZCF). They had a blurred idea that the cost of mealie meal was being fixed by millers. The unfortunate thing is that even after installation of these meals and having two consecutive good yields the price of mealie meal has not gone down.

With five acres, what can one grow?

"I have five acres of land, and what can I profitably grow on it"? This is a summary of a text message I received from a forty-year-old Zambian would be farmer who was working as a marketer in 2012. From the tone of the text message, one could tell that she was a lady and who later came to be known as Ms. Mulenga (not real name). Ms. Mulenga had acquired a five-acre land in Lusaka west and she shared that she had passion for farming. She was contemplating switching her career in marketer to go into full time farming but did not have enough capital and knowledge on enterprises to ventured into. Five acres of land is equivalent to two hectares of land. Basically, Ms. Mulenga could do a lot on her land and she could grow any type of crop on that land.

However, certain crops would give more returns than others on the same unit on land. For instance, it would be folly for her to quit her job and venture into maize production on a two-hectare piece of land and think it would sustain her. The lady could be successful but will not make the money that she wants to or equivalent to what she was earning in formal employment. With a yield of five tons per hectare, which is generally the best average yield for most of the small scale farmers in Zambia, this will get her two hundred bags of maize. At the FRA prices then, she would get K14, 000 against minimum cost of inputs of about K9, 664 assuming she uses herbicides, fungicides, seed and fertilizer. This will make her remain with a gross profit of about K4, 000 for the whole year. However, using the same piece of land and growing green maize (assuming she has irrigation) would give her a gross income of about K88,000 (assuming that all the cobs are sold at K1). With this income, she can boast that she is a farmer worth calling one.

The upcoming farmer needs to do a thorough planning of what she wants to venture into to help her make informed decisions. People ought to know that agriculture is not all about growing crops, others practice agriculture by rearing animals and birds. For such limited piece of land, the best gamble is for her to go into mixed farming where she could be rearing birds at the same time growing some vegetables like tomato, cabbage, pumpkin leaves (*chibwabwa*) and other horticulture crops. Potato production too, is an enterprise that is on the upswing and with the opening of fast foods and malls at every corner in Lusaka, venturing into growing the potato crop in Zambia will not be a wrong idea to think of.

In the earlier articles in this book, it was shared with how effective has been production at one of the farms in Lusaka called York Farm. This was a farm that was established in the early 80s (1978 to 1982) on a four-hundred-hectare piece of land on which production goes on for 365 days in a year. In my last year of my attainment of an agriculture degree, I happen to have worked on that farm and was appalled at the level of intensive agriculture that goes on. The first instinct, being a student of soil science then was that the heavy machinery they were using could been causing subsurface soil compaction. This prompted me to do a final year research for my dissertation to find out the effectiveness of the production practices done to avert soil compaction, and the results were negative – no soil compaction was caused! This comparison I am trying to bore the readers with is an indication that regardless of how small the land is, a farmer worth his salt can do something that can sustain the farm and have a meaningful livelihood better than even those that are in formal employment. In Zambia today and Africa in general, there are farmers that are surviving on very small pieces of land but are well off than some of the farmers that have huge chunks of land. Some may not believe that on that five-acre piece of land, they can rear three daily cows that can give them milk every day. Indeed, there are so many things that can be done on a five-acre piece of land, but what the farmer first needs to do is to understand the land capability by asking qualified officers from the ministry of agriculture to conduct land capability and suitability assessments. Some lands may look suitable for arable farming when the soil depth is only 20cm or less, and that will let one know what crops can be grown on such land.

This takes me to the next plea to some of the town dwellers who thinks farming can only be done in villages. I have admired some yards in some compounds especially in low density areas like Kabulonga, Woodlands, Chilanga and many others. Some inhabitants in those areas have yards with provisions of gardens that they have left just like that. Some of them have ended up planting flowers that are not edible; yet everyday they trek to markets, Shoprite and other supermarkets to buy basic things like cabbage, onion and other vegetables. There is no justification for anyone to be buying cabbages for instance if one has about twenty flower pots around the house and they have constant water supply. One can grow at least one head of cabbage in each of the flower pot depending on the size and this will reduce the cost of them buying vegetables. The beauty of us practicing such types of agriculture is that it helps us develop the farming skills by understanding diseases and certain crop requirements such that even the time we are thinking of retiring to our villages (as it is the norm for most of you who think farming is for the tired finished people), we will know how it is done. Everyone need to pluck out all those weeds they have in their flower pots and plant something which will help them save money. Five acres of land starts with a couple of flower pots.

What a huge potential for the livestock sector!

In this country, when people talk about agriculture they normally refer to crop production and in most cases it is field crops. However, agriculture comprises of many facets including livestock production. This sector has great potential, more than what we can realise from field crop production. This can be demonstrated with an example. Apart from the tomato smallholder farmers in Zambia, there are very few farmers in this country can walk into Toyota Showroom to buy a brand-new vehicle? Having driven and interacted with smallholder farmers from the breadth and length of this country and I can hardly remember any such that has bought such an asset through maize production. Many of us rush for second-hand vehicles from Japan the time we want to buy any. It is normally not our choice that we buy such vehicles but we are limited by the resources we make from our field crop production businesses. Nonetheless, I can share with the readers that I have seen several of the cattle herders from Namwala district buying very good brand-new cars including the famous Mercedes Benz. What is funny is that people who don't appreciate the value of livestock go to an extent of laughing at such farmers.

In this part of the world, we think agriculture can only be practiced if one has fertilizer to grow the low-value crops like maize; it is identifying the enterprise that will give you more value and developing a competitive edge in producing such commodity. The Tongas of Namwala and other areas of Southern Province have realized that the weather pattern is no longer favouring them and they can get more value from rearing cattle than growing field crops. Of the five million head of cattle that are in Zambia, half of that is found in Southern province. On average, a cow can take three to four years to reach market weight and a good-looking cow can fetch as much as four thousand (US$400) per head. Some people in Southern province have as much as five thousand heads per person. If such a person decides to sell about three hundred animals, he will realise a value of K1.2million.

A Toyota Hilux double cab motor vehicle from Toyota Zambia costs about US$50,000 which in our currency is approximately K500,000 at the June 2018 exchange rate (US$1=K10). Such a farmer can easily buy a brand-new Hilux and remain with K700,000 the money which can build himself a good house if not two like the ones that ministers lives in, send his children to very good schools and support other livelihood daily activities. In the local language, the people that look after cattle are called 'Kachema' or herders! This to some can be derogatory if they do not know the value of livestock. Zambia's land can support as much as fifteen million cattle and the epidemic diseases that keep breaking out so often can be contained, Zambia can be supplying lucrative markets such as the EU like our neighbor Botswana does. By exporting the meat products, the country will be creating more sustainable jobs and improving the economies of the farmers in particular and Zambia in general. I have always said the vast resource we have in land in this country, if well used, can make this country even forget about burrowing in the ground searching for copper that we don't own as the mines are now owned by foreigners.

There is no country in this region which is blessed like Zambia is. Going forward, what do we need to do as a people? The production of livestock need to be segmented into three areas. Southern and Western provinces can be segmented into one region. Central, Lusaka, Eastern and Copperbelt provinces can be another region which are relatively stocked with animals and the last region would be Luapula, Northwestern, Mchinga and Northern provinces which are sparsely populated with animals. In this regard I feel that there is less that I feel can be done for Southern Province as far as cattle rearing is concerned. However, there is need to increase the number of watering points such as dams so that the animals don't mingle with buffaloes that carry some of these diseases as they stray from national parks. The country also needs to build more dipping infrastructure to control disease causing pests such as ticks and pathogens. In addition, farmers need to be trained and sensitized on feed formulation for feed supplementation which is normally a problem in drier seasons. These activities however, can be replicated in other provinces, which have relatively fair population of livestock. For the moderately stocked regions, there is need to introduce improved breeds to interbreed with the Angoni cattle, which have a 'short wheel base' to improve on the size and weight. There is also need to rid of the types of pigs found in Eastern province with pure breeds such as the land lace or other good meat, producing breeds. On the other hand, we need to do a bit more than just introducing pure breeds and controlling diseases in Northern, Muchinga and Luapula provinces. This is because of the cultural backgrounds of the people from these regions; they are traditionally capture fishermen and rarely rear livestock. Their source of protein is mostly from fish and beans. They do not have time to go and herd their livestock but would rather go fishing in the various water bodies and sometimes they go hunting for smaller wild animals. Firstly, it should be demonstrated that they will get more value from rearing pigs, cattle, goats, chickens and keeping fish than venturing into fishing in lakes such as lake Bangweulu, which has more crocodiles than fish due to over fishing. The fish in the water bodies from Luapula has been depleted and people have ended up trekking south to lakes such as Itezhi-Tezhi and Lake Kariba to fish. There should be campaigns to sensitize our brothers that they will get more benefit from the other enterprises such as rearing cattle in the vast sweet grasslands in their lands. Once they understand the benefit and with a lot of exchange visits to areas such as the Southern, only then can we think of restocking these provinces. In a nutshell, we have a greater opportunity to turn around this sector for economic prosperity than growing maize. There is more value in 'cow dung' than in maize Stover. Nonetheless, the government has done well by splitting ministries of livestock from crop agriculture. This should be followed by timely allocation and release resources to this ministry.

By putting up good policies and implementing the right activities, agribusiness in this country can be as lucrative business as is the motor industry in Japan. We know that at independence, Zambia's population was just over three million. The country had more assets than the population and each person could afford to own land not less than 0.25 square meters. The amount of trees, fish, wild animals, minerals and other assets was immense. As mentioned in my previous chapters, our economy was equivalent to, if not better than, that of South Korea. Currently, because of poor management of these resources coupled with bad leadership that is greedy, we have destroyed these important assets to the point that we all look forward to an opportunity to leaving this country for others.

However, not all hope is lost if we can come back to our senses and manage what is remaining. In the 70s, Zambia was known for its copper and that was the industry where it used to get its resources. Currently, all the mines are in the hands of foreigners or foreign companies. Little is known of how much these companies are making because all the money realized from the proceeds of minerals is banked in offshore accounts. We only remain with destruction like the Sulphur dioxide pollution in Mufulira, the lead contamination in Kabwe and so many *filongoma* (trenches left behind) like those in Chingola. I have come to learn that we are a people that don't love themselves. We have fallen for destruction of our own environment in a corrupt way never seen before. Much as this is the status quo, we are remaining with one asset and that is land. Of all the factors of production, land is the most important. Our land is so fertile with abundant fresh water for irrigation with good sweet vegetation that can support production of different livestock. Our forests are endowed with the best forests products such as the mukula tree and honey. This is the last resource we are remaining with and the most important. Therefore, we need to manage it so prudently with sober minds. If we have to dig deep in history, we will find out that both the world wars were fought because of land. The current problems that are in Syria are not because people want to be presidents but it is because of land which is rich in oil and uranium if not mistaken.

Our land, just like in other many countries in Africa has oil, uranium, diamonds, gold, cobalt and rich soil nutrients. People are travelling thousands of miles to come and buy our land not that there is no land where they are coming from; they have heard how our land is rich. As a people, we need to be careful of how we dispose of this land to foreigners at the expense of indigenous Zambians. No country will be a colony of another like it was in the 1800s but colonization has now taken a new twist. We cannot talk of the economy in this country but we can raise our heads high when talking about land. We have enough time to make amends for the mistakes we made over proceeds of copper to administer land with sober minds. There is need to develop a land policy which favors redistribution of the land to the local people. This is the capital which we the locals can use to buy shares in companies.

China, India, USA, UK and many others although more developed than us, they also have people that are homeless and live on less than a dollar a day. Let us open our eyes, sober up and be patriotic enough not to waste our land. We are lucky to be found in this part of Africa but if our egos override the principle of being Zambians, we will have ourselves to blame just the same way we are blaming each other over the poor drainage in Lusaka. We should for once stop this indiscriminate sale of land. Again, I emphasize that land is the only asset we are remaining with as a country and we need to manage it with caution. Land will not rot, so we shouldn't be in a hurry to dispose it off.

Potato, an alternative crop to tomato

In first John 2:17 the Bible gives an assurance that the world is passing away and so are its desires, but the one who dose the will of God will remain forever. Those that lived thirty years ago and are still living to date will agree with the Bible because of what is happening and many things are being fulfilled. Twenty years ago, it was very difficult for many farmers to cultivate tomato and there were very few farmers growing the crop. Today, there are so many farmers growing it and this has sometimes led to the crop going to waste because there is no supporting marketing infrastructure to anchor activities beyond production. These are activities which will promote value addition to increase the shelf life of the product and specialized vegetable markets like is found at Johannesburg Fresh market in South Africa. Farmers are still mourning the low commodity prices for tomato. There are still some more fruits to be brought to the two key markets in Zambia; Soweto and Chisokone. This trend will continue for many years and hopefully throughout many production seasons. In mid-2018, there were reports that the country might experience a drought or longer dry spells; drier than what we experienced last production season, and this could work to some farmers' advantages. The fact is that farmers will have to be in production regardless of what happens, besides the mines which were our main industry, are in foreign hands who are busy externalizing the money back to their countries. Zambian farmers and smallholders in particular have learned the art of growing tomatoes. They are producing quality tomatoes that can sell anywhere on the planet. Now that many of them are all growing good crop of tomato, what else can they diversify into as farmers?

Tomato belong to the family of Solanaceae and some other crops that belong to this family are potato, chilli egg plants, pepper, tobacco and others. When the environment for these agricultural products is scanned, one will agree that Zambia has more shopping malls than manufacturing plants. These malls have shops selling clothes from China and South Africa as well as fast foods or 'junky' foods as they are now called. These fast foods will need tomato but not as much as they will need potatoes for making French fries. This is an opportunity that tomato growers have and can easily tap into. Those that have grown tomato can easily grow potatoes because the diseases and pests that attack both crops are similar. This is an opportunity some of the tomato growers can diversify into off the rainy season when they are not growing maize. The advantage of the potato crop is that it has a longer shelf life than tomato. Potatoes are mainly grown from tubers and these tuber seeds can easily be bought from Buya Bamba in Lusaka. The choice of the seed variety is very important because some are more prone to diseases than others. The size of the tuber is also critical because the size determines the number of stems it will have. In addition, a larger tuber will have more eyes and stores more carbohydrates, nutrients and water which are critical to sustain the young seedling. The crop is grown on ridges for easy harvesting and the planting of the seeds can start as early as March in this part of the globe. The planting and specifications of the ridges will depend on the variety and size of tuber seeds.

Potatoes should be grown in soils that are friable and easy to work with and they prefer pH range of 5.5 to 7. A pH that is too acidic will render certain nutrients especially the macro ones to be unavailable. One of the key nutrient to watch for under acidic condition is phosphorus which will be unavailable or locked as we call it in agronomy. A pH range of 6 to 7 promotes the availability of most plant nutrients while on the other end, very high pH of over eight (8) will limit the availability of iron, magnesium and boron. Knowing your soil status is not only important in cultivation of potatoes but other crops as well. It is also important that the soil should contain adequate amounts of organic matter as this acts as a reservoir of carbon and helps to store water as well. Potato need a higher requirement for phosphorus, potassium and nitrogen. These should be readily available as you may know that these are also important for protein formation as the crop is high in protein content. It also requires some secondary elements like magnesium which is an important constituent of chlorophyll for photosynthesis, calcium and Sulphur but not as important as the micronutrients such as manganese, iron, boron, zinc, and copper including molybdenum. This is the reason why it is important to apply foliar fertilizers in the production of potatoes because these might be locked in the soil especially if we are not monitoring our soil nutrition. This is a crop that has a highest susceptibility to phosphorus deficiency. This can be seen on the leaf as well as root size, ultimately affecting the tubers size, quality and number of tubers formed. This element is only second to nitrogen in the effective growth of a potato plant. However, it is important to note that no nutrient is more important than the other. The absence of one nutrient either micro or macro will be the limiting factor in the effective production of the crop and it is the reason that farm managers should closely monitor the crop nutrition.

The second aspect of the effective production of potatoes just like tomatoes is managing the diseases and pests. One of the most devastating disease for potato is early blight. This can easily wipe out the field if not controlled. The only good news that we have is that there are a couple of products that can be used to control this and one of the products I have come to like both for tomato and potatoes is Bellis. This is a must have product. The other most important disease which caused famine in Ireland and killed over a million people in 1800 is the dreaded late blight. It is a fungal disease caused by a group of fungi called the Oomycetes. Normally in the growth of potatoes just like tomatoes, the late blight sets in early and the other one late but it doesn't mean the opposite cannot happen. Again there are so many products that can control these diseases but what is important is to spray preventatively as the saying prevention is better than cure. Other diseases of economic importance are black rot, brown spot, fusarium wilt, powdery mildew and scab, verticillium wilt and other viral diseases. It is important to note that just like in humans there is no pesticide that can control viral disease. What is important is to control the vectors which carry and transmit these diseases such as aphids and grasshoppers. On the pests, look out for the dreadful red spider mites especially now that it is hot and whiteflies, potato tuber moth, the leaf miner and the nematodes. On the other hand, you will agree with me that marketing of potatoes is slightly easier than tomatoes because of the shelf life. My fellow farmers, let's not cry about the loss we have incurred from tomato; that is how business is, you win some and lose some. What is important is to soldier on whether FISP is there or not. Zambian farmers are so resilient and I know that they will produce even without FISP. Mind you, the minister of agriculture does not know when the inputs will be distributed and in less than six weeks, farmers will be planting. How I wish he salary can be withheld until he assures the farmer of the date when the fertilizer will be distributed.

The sustainable management of natural resources for tourism

When the first people came to Zambia, they should have enjoyed the serenity of the environment around them. They never used to cultivate but were sourcing their food by collecting fruits and hunting from the forests; there was no shortage of what to eat, and the air they breathed was fresh and full of clean oxygen. Our environment today is totally different to what it used to be; it is becoming inherently harsher and unfriendly to live in than what is has been. Human beings have destroyed the environment in the quest to seek better living through economic and social development. Countries like Zambia had very good forests covering most landmass of the country which were able to recycle the air they breathed. Nowadays, its inhabitants are made to breathe in toxic elements such as Sulphur dioxide and carbon monoxide contaminated air as is the case in Mufulira's Kankoyo township. The people of Makululu, Kasanda and Chowa in Kabwe are exposed to lead, a very dangerous substance which is so harmful to their health. Just recently, some communities in Chingola were made to drink contaminated water which caused some complications to many people. Many communities in Southern province have challenges of accessing water especially in the dry season because most of the trees have been cut down and disturbed some watersheds which has led to erratic rains and drying up of some perennial streams. The litany of problems goes on and they all point to the unsustainable activities of man in the quest exploit the natural resources for a better living. These problems are not only synonymous to Zambians but it is even worse in some highly developed countries, where it is almost impossible to breath freely in certain towns because of air pollution.

Should the people continue with their destructive acts in the name of development, the environment will even become more unbearable. Though the country wants to develop, it should not come at the cost of the environment because then such development is not sustainable. As individuals, it is time each one reflected and ensured that all developmental activities that are undertaken are done in a sustainable way and environmental friendly manner. Not long ago, it was rare to hear people mention climate change; it was not in our vocabulary and probably that could be the reason certain people believe climate change is a hoax. Peoples livelihood is threatened by this phenomenon of climate change; and if people don't change their habits, the environment will be destroyed to a point of not reversible. In a country like Zambia, people risk being decimated because of the impact of climate change and the biggest risk will be shortages of water and ultimately food. If Zambia as a country can have three to four seasons of droughty conditions, there will be great shortages of food and water and animals, especially wild life and livestock will die. Therefore, it is prudent upon every one of its inhabitants to effectively conduct their daily economic activities with sustainability in mind. Many of people think sustainable development only applies in agriculture, mining and such macro-economic activities. Not at all, it starts with small things that people do in their daily lives; things such as throwing liter anyhow, throwing away left over foods, washing their cars using horse pipes, leaving their security lights on during the day when they are away from home, including the deadly effects of corruption. It is prudent therefore, to manage natural resources including wildlife so prudently. Tourism especially wildlife tourism is going to be a very big business in the near future. This is because many animals are becoming extinct due to the destruction or clearing of forests for agriculture.

One area we can easily mitigate the effects of climate change is by maintaining health green forests and a balanced fauna. This can be done by stopping the indiscriminate cutting down of trees for instance; especially trees meant for charcoal burning and tobacco curing. How can the tree cutting for charcoal burning be stopped? Forestry management is one area we have fared so badly as a country. In our lovely country, it will be unthinkable to tell people to stop cutting down trees if they are not made to understand why they should not do it. Many of them are cutting trees as a source of energy for cooking. There are many reasons for this but one of them is lack of access to electricity. Only about 3.1% of the rural areas are electrified or have access to electricity. This means about 97% of the rural communities are using either charcoal or firewood as sources of energy. It is not feasible and possible at the same time to tell such people to stop using forest products such as wood for energy. The first step is to ensure that they are provided with alternative source of energy through connecting them to the national grid for tapping to electricity for instance. Alternatively, there are so many solar energy products and equipment that can be promoted in the rural areas. This is a great opportunity for investment in solar products. Strides such as those being promoted by SNV and Oxfam to promote biogas in rural and peri-urban areas should be supported and must have a budget line by the ministry of environment as well as the private sector. Secondly, other people cut trees for charcoal burning because that is a source of income for them. To stop such people from engaging in such destructive activities, need concerted efforts. They need to be empowered with alternative livelihoods such as farming and agro-tourism for instance. Many farmers have had challenges with production for many reasons. Others could have produced but lack of access to markets for their output makes them make less or no money at all. Once these basics are put in place, only then can such people be partners in forest management. Many people have taken up rearing of wildlife for domestic tourism and they are making good money from such activities.

Therefore, stopping people from indiscriminate cutting down or destroying forests should not be a preserve of the forest officers; this is for everybody. As a country, it should embrace technologies that can help with management of forests. The use of drones in monitoring the forests is one technology that should be adopted. Authorities are aware that people are out there cutting down *mukula* tree and the forest departments are ill equipped with transport to combat such vices. The use of drones to monitor what activities are being undertaken in forests will go a long way in reducing the scourge of where forests are depleting at a faster rate than the planting of the same. Using of the satellite imagery is another opportunity that needs to be explored. Maintaining forest cover will mitigate climate change and it will greatly contribute to the tourism in particular and economic development in general. This country is lucky that there are still some forests and wildlife but management of these natural resources should be stepped up, especially now that we are experiencing adverse impacts of climate change on our environment.

ACRONYMS

ACC	Anti-Corruption Commission
BASF	Badische Aniline und Soda Fabrik
BOZ	Bank of Zambia
BSc	Bachelor of Sciences
Ca	Calcium
CAC	Camp Agricultural Committee
CBD	Central Business District
CDC	Commonwealth Development Corporation
CEEC	Citizens Economic Empowerment Commission
CEO	Chief Executive Officer
COMESA	Common Market for East and Southern Africa
CFU	Conservation Farming Unit
CNN	Cable News Network
Cu	Copper
DC	District Commissioner
DMDC	Diocese of Mongu Development Committee
DMMU	Disaster Management and Mitigation Unity
DPB	Dairy Produce Board
DRC	Democratic Republic of Congo
EXIM	Export and Import
EU	European Union
eVoucher	electronic voucher

FAO	Food and Agriculture Organization
FDI	Foreign Direct Investment
Fe	Iron/Ferrous
FISP	Fertilizer Input Support Program
FOREX	Foreign Exchange
FRA	Food Reserve Agency
GART	Golden Valley Agricultural Research Trust
GBM	Geoffrey Bwalya Mwamba
GDP	Gross Domestic Product
GMAs	Game Management Areas
GTAZ	Grain Traders Association of Zambia
H	Hydrogen
Ha	Hectare
HIV/AIDS	Human Immune Virus/Acquired Immune Deficiency Syndrome
HH	Hakainde Hichilema
HP	Heritage Party
INDECO	Industrial Development Corporation
K	Potassium/Kalium
Kg	Kilo gram
MA	Masters of Arts
MAL	Ministry of Agriculture and Livestock
MBA	Masters of Business Administration
MDG	Millennium Development Goals
Mg	Magnesium
MMD	Movement for Multiparty Democracy
MP	Member of Parliament
MRI	Maize Research Institute
MSc	Masters of Sciences
Mt	Mount/Metric ton
N	Nitrogen
NAMBoard	National Agriculture Marketing Board

NAP	National Agriculture Plan
NERICA	N Rice for Africa
NGO	Non-Government Organization
NRDC	Natural Resources Development College
P	Phosphorous
PAM	Program Against Malnutrition
PF	Patriotic Front
pH	negative log Hydrogen/measure of acidity
PTC	Post and Tele Communications
ppb	parts per billion
PLARD	Programme for Luapula Agricultural and Rural Development
QDS	Quality Declared Seed
RABO	Dutch Bank
ROP	Refined Oil Products
R&D	Research and Development
RFP	Request for Proposals
S	Sulphur
SADC	Southern Africa Development Community
SILC	Savings and Internal Lending Communities
SME	Small and Medium Enterprises
UBZ	United Bus of Zambia
USA	United States of America
USAID	United States Aid
UNZA	University of Zambia
UK	United Kingdom
UNIP	United National Independence Party
WCS	Wildlife Conservation Society
WRS	Warehouse Receipt Systems
ZAWA	Zambia Wildlife Authority
ZAMACE	Zambia Agricultural Marketing Commodity Exchange
ZAMEFA	Zambia Metal Fabricators

ZARI	Zambia Agricultural Research Institute
ZEMA	Zambia Environmental Management Authority
ZCF	Zambia Cooperative Federation
ZNBC TV	Zambia National Broadcasting Corporation Television
ZNFU	Zambia National Farmers Union
Zn	Zinc
ZESCO	Zambia Electricity Supply Corporation
ZCBC	Zambia Consumer Buying Corporation

ABOUT THE AUTHOR

Felix Tembo is a seasoned agribusiness development expert and a business management practitioner. He has over fifteen years of experience and has held senior management positions in multinational companies at different levels in sales, technical sales and operational management. He has also worked in International Non-Governmental Organisations (NGO) as an Agribusiness Specialist in project management at consultancy and advisory levels. He has passion in agribusiness and would want to see agricultural productivity improved especially for smallholder farmers through development of sustainable markets so that they can operate as real businesses which will stir up production. As a patriotic Zambian, he has worked for multinational companies mostly in Zambia and Southern Africa with companies such as Syngenta, BASF, CropChem Services and on project run by international NGOs such as TechnoServe, SNV Netherlands Development Organisation, NIRAS Finland as well as USAID funded project. He has done several consultancies for international companies and NGOs in the area of agribusiness especially in market development (M4P) and productivity improvements for smallholder and semi commercial farmers using inclusive business models. He is a holder of a Bachelor of Sciences degree in agriculture from the University of

Zambia and MBA from Management College of Southern Africa. He also holds a certificate in Sales and Marketing from Zambia Open University. Currently, Felix is pursuing a PhD in Management with some American University. He is working as a Technical Sales Support Manager for Southern Africa as well as Country Manager for Malawi with one of the multinational company.

ABOUT THE BOOK

Zambia just like many countries in Africa has faced several challenges in its quest to attain economic and social development. One sector that the country has endeavored use as a driver for for development agribusiness. Agribusiness besides mining is the oldest and significant industry to Zambia's foreign exchange earnings. It contributes ernomously the country's gross domestic product (GDP). Since the slump in the copper prices, production of copper went down leading to declined exports. This has resulted in reduced forex inflows from the mining as it is the major contributor to the economy. The decline in mining activities has provided opportunities for the agribusiness sector. This has become one of the key sectors and it is also the largest provider of employment opportunities in the country, contributing over seventy percent of jobs in the country.

This book profiles the progression of agribusiness development in the region and Zambia in particular. It brings out some of the salient points that have led to agribusiness development, the enablers in the sector and the comparative advantages that the country has in the region. It highlights some reasons which led to stagnation in the initial stages of the sectors development. It also highlights some of the favourable policies the country has formulated and how they have contributed to it attaining a competitive edge in Southern Africa. It emphasizes the fact that the agriculture sector can be a cornerstone of the manufacturing sector. The book also highlights some of the great potential the country has and some other sectors that have great opportunities for investments, such as energy. For those that are thinking of investing in Africa and Zambia in particular, this book can serve as a quick guide. The book clearly points out the key sectors as being agriculture, tourism, mining, energy, trade and manufacturing as well as construction and the services industry. In the closing chapter, the book draws the readers to some of the articles that were published with three local tabloids from 2012 to 2018 to give it a practical touch to agribusiness and some of the successes achieved in the country thus far.

Made in the USA
Middletown, DE
05 February 2022